the
Experimentation Machine

the
Experimentation Machine

*Finding
Product–Market Fit
in the Age of AI*

Jeffrey J. Bussgang

Copyright © 2024 Jeffrey J. Bussgang

All rights reserved.

No portion of this book may be reproduced in any form without permission from the publisher, except as permitted by U.S. copyright law. Every effort has been made by the author and publishing house to ensure that the information contained in this book was correct as of press time. The author and publishing house hereby disclaim and do not assume liability for any injury, loss, damage, or disruption caused by errors or omissions, regardless of whether any errors or omissions result from negligence, accident, or any other cause. Readers are encouraged to verify any information contained in this book prior to taking any action on the information.

 Published by Damn Gravity Media LLC, Chicago
www.damngravity.com

ISBNs:
Hardcover: 978-1-962339-08-7
Ebook: 978-1-962339-09-4
Audiobook: 978-1-962339-10-0

Printed in The United States of America

Table of Contents

Introduction	Timeless Methods, Timely Tools	vii
Chapter 1	The Science of Startups: How to Build an Experimentation Machine	1
Chapter 2	The 10x Founder: Harnessing AI to Create Startup Superpowers	17
Chapter 3	Having a HUNCH: Redefining Product-Market Fit	35
Chapter 4	Hair on Fire: Discovering Your Customer Value Proposition	55
Chapter 5	Laws of Attention: Go-to-Market in the Age of AI	85
Chapter 6	The 10x Club: Building a Magical Business Model	115
Chapter 7	Scaling Product: Growing Beyond Product-Market Fit	137
Chapter 8	Scaling Go-To-Market: Sales, Marketing and Expanding TAM	153
Chapter 9	Scaling People: Hiring in the Age of AI	175
Chapter 10	Responsible AI: Embracing Ethics, Diversity, and Social Equity	187
Conclusion	The Experimentation Mindset	211
Appendix A	AI Prompt Tips and Examples	219
Appendix B	Startup Valuations	237
Appendix C	Cohort Analysis Examples and Explanation	253
Appendix D	a16z's AI Apps Unwrapped	265
Endnotes		269
Acknowledgements		277
About the Author		279

Introduction

Timeless Methods, Timely Tools

THERE'S A RUNNING JOKE in Startupland[1] that artificial intelligence (AI) is going to replace *all* humans inside a business: the founders, product managers, salespeople, engineers, and certainly the lawyers. While some tasks will be automated, I don't expect fully autonomous startups anytime soon. But massive changes *are* happening, and some folks will be left behind.

AI may not replace the startup founder, but startup founders who use AI are going to replace founders who don't.

My fascination with AI began as a computer science major at Harvard University in the late 1980s and early 1990s. That era was marked by the birth of personal computers and the emergence of telecommunications protocols that enabled the widespread use of email and eventually provided the underpinnings for the Internet. It was a time of rapid technological advancement, and I was captivated by the potential of these new tools. As a science fiction buff (Isaac Asimov, in particular), I loved to imagine a world of robots and cognitive aids. I took graduate courses in neural networks, natural language processing, and computer vision in parallel to my undergraduate computer science work. My thesis was titled *Coordinating Natural Language and Graphics in the Automatic Generation of Explanations of Causal Business Models*—a long-winded way of saying I applied early AI concepts to generate sentences that explained business decisions. Sound familiar?

After completing my computer science degree and then later my MBA from Harvard, I entered Startupland, where I learned how to apply leading-edge AI in practical use cases. As an entrepreneur, I joined and later co-founded software companies that harnessed early AI and machine learning technology. One of these ventures aimed to personalize e-commerce experiences (Open Market, IPO 1996) and another helped parents and students save for college (Upromise, acquired by Sallie Mae). Since co-founding the venture capital (VC) firm, Flybridge, I have led the seed rounds for several groundbreaking AI startups, including one that utilized machine learning to provide fair and transparent underwriting of consumer loans (Zest AI) and another that created an early robot companion for the home.

When ChatGPT debuted in late 2022, I knew we had reached a point of no return.

Generative AI—powered by Large Language Models (LLMs) to create strikingly creative and human content in text, code, and visuals—is an incredible example of science fiction becoming science fact. From the consumer's perspective, computers suddenly became capable of interacting in ways so humanlike that the difference between human and machine was imperceptible. They could tell you bedtime stories, generate cover letters, code up a website, and even paint you a Monet. The famous Turing Test—a capabilities threshold to ascertain if machines could pass as human—was shattered.

Gen AI's rate of adoption is nearly as staggering as its creative output. It took the World Wide Web seven years to reach 100 million users. ChatGPT reached that milestone in just two months. As of this writing, ChatGPT (still the largest AI platform) has over 300 million *weekly* active users in just two years since its launch, and it continues to grow by double digits every quarter.

For businesses, generative AI represents the largest boost in productivity and opportunity since the Internet. I have become fascinated with how to apply this radical new technology in startups. AI has become a focal point in my Harvard Business School (HBS) MBA course, Launching Tech Ventures. We've also gone all in on AI at Flybridge, rewriting our investment thesis to exclusively focus on "AI-forward" startups across industries.

In particular, I am interested in how startup founders and their teams can use AI to accelerate the search for product-market fit, and later, to scale. Like a catalyst in a chemical reaction, AI tools have the power to turbocharge startups through the learning and growth curves, creating bigger and better outcomes, faster than we ever thought possible. *Founders who use AI will replace founders who don't.*

But ChatGPT and other AI tools are just that—tools in the startup toolbox. Founders still need to answer the same fundamental questions they always have:

- Who is my customer?
- What problem can I help them solve?
- How do I make money and build a sustainable, profitable business?

To answer these questions, I have long advocated for a scientific approach to building startups. Founders should create hypotheses, run experiments, and implement the results. That's why my preferred mental model for startups is the *Experimentation Machine*. The goal as an early-stage founder is to maximize your learning by testing assumptions and refining your ideas. The key is to do this quickly and successfully enough to find product-market fit before you run out of money. AI is the ultimate experiment accelerator for startups.

This book will guide you through the process of building your own Experimentation Machine in the age of AI. I recognize I am trying to hit a moving target when it comes to describing the capabilities of this technology. The field is advancing at a dizzying pace, but I aim to give you a tried-and-true playbook for finding product-market fit while utilizing the most cutting-edge tools available today—no matter what field you are operating in.

Here is my goal for this book: *Timeless methods. Timely tools.* We will study companies from before and after gen AI's

"Big Bang" moment and apply the timeless methods of the Experimentation Machine methodology while pushing the envelope with modern AI-powered strategies. In the chapters that follow, I will draw from my HBS case studies, as well as Flybridge portfolio companies, to lay out the playbook to methodically transform your startup into an experimentation machine, catalyzed by AI, to efficiently find product-market fit and build a successful, enduring business. Specifically:

Chapter 1 breaks down the Experimentation Machine methodology. You will learn the fundamentals of building an experiment-driven organization across product, growth, sales, and business development. You will discover why test selection is the core component of startup strategy and how to sequence your experiments, studying the case of Squire.

Chapter 2 explores the concept of becoming a 10x Founder: a new generation of super-founder who harnesses AI to supercharge their experimentation, leverages their extremely limited assets, and achieves rapid scale *without* organizational bloat. To illustrate, I'll introduce you to the founders of Topline Pro and Talktastic.

In Chapter 3, we will discover a new definition of product-market fit (PMF)—one that looks beyond growth metrics to ensure you're building a sustainable, long-term business. We will cover the HUNCH framework for evaluating the strength of your product-market fit and look at startups like Superhuman and ClassPass as examples.

Chapter 4 covers the first set of experiments every startup must run: refining your customer value proposition (CVP). You'll learn to generate and test hypotheses to refine your product offering and how AI can help in that process. The innovative biotech startup, C16 Biosciences, will serve as a case study, along with revisiting Squire.

Go-to-market strategy (GTM) will be the focus of Chapter 5, including tests to discover your ideal beachhead market, sales model, and channels. We will also explore how AI-forward startups are supercharging their GTM strategies. This chapter will feature case studies of Ovia and Shippo as well as our portfolio companies AllSpice and Teal.

Chapter 6 will introduce you to the "magical" business models that yield the highest valuations. We will cover the critical importance of unit economics and the dangers of delaying revenue for too long. You'll see how companies like Khatabook and Ovia launched experiments to help find their winning profit formulas.

Chapter 7 will introduce the power of AI agents to scale your product from a Minimal Viable Product (MVP) to a robust offering in the age of AI. I will share some specific no-code and low-code strategies to build faster, iterate better, and avoid technical debt. Case studies featured include Blitzy.AI and Shippo.

In Chapter 8, we will cover strategies for scaling go-to-market and expanding your total addressable market (TAM). I will introduce the Sales Learning Curve, AI-powered marketing tactics, and three ways to reach new markets.

Chapter 9 will focus on scaling your people in the age of AI. We will discuss who to hire (introducing you to the 10x Joiner), *how* to hire, and when. Since AI is making startups more efficient, every new employee matters more than it did in the past.

Chapter 10 shifts our focus to the risks and challenges that AI presents, as well as long-standing systemic issues in Startupland. We will first look at AI-specific risks, like the environmental impact of compute, and then tackle the challenges of ethics and equity in the startup ecosystem, because AI will only exacerbate existing issues. You'll learn how to navigate your organizational biases and build an inclusive company that makes a positive impact. Featured case studies include Juul and Snapchat.

In the Conclusion, I will say a few final words on the experimentation mindset that founders must acquire to be successful today. This mindset embraces continuous learning, AI leverage, and giving back to the Startupland ecosystem that helped make you successful.

Last but not least, I share four deep dives in the Appendix on important topics that didn't quite fit into the book's main narrative: AI Prompting Tips, Startup Valuations, Cohort Analysis, and a curated list of AI tools.

While I will mention a few specific, state-of-the-art AI tools in this book, there are many cases where I refer to "AI tools" generally because, frankly, there are dozens of options, and the best ones will likely be different when you're reading this than the day I'm writing it. When I reference an AI tool for sales enablement, for instance, I leave it to you to research the best product available. My goal is to simply make you aware of all the possibilities at your fingertips.

Let me share one last piece of advice: You may feel compelled to wait to adopt AI in your business until it gets better, more mature, or more precise. But remember that the AI tools you use today are the *worst* they will ever be. If you wait until AI is "perfect," you will be waiting a long time, and by then it will be too late.

As futurist and sci-fi author William Gibson said, "The future is here, it's just not evenly distributed." This is your chance to live in the future and reap the rewards. I hope this book will guide you on your journey through Startupland to great success. Good luck!

—Jeff

Chapter 1

The Science of Startups: How to Build an Experimentation Machine

EVERY SCIENTIST NEEDS A lab. For Songe LaRon and Dave Salvant, the founders of Squire Technologies,[2] that lab was a barber shop in Manhattan's Chelsea Market.

LaRon and Salvant started Squire with a simple goal: make it easier to book a haircut. "Every time you needed to get a haircut, it was unpleasant," Salvant said. "You either had to plan ahead and call to make an appointment, or you had to show up in the shop and wait for sometimes an hour or more; you never knew how long it would be. You needed to get cash because most barbers were cash only. You walk out feeling great, but it was painful to get there."

The duo imagined building the OpenTable for barbershops—an app that would allow patrons to schedule haircuts at their convenience. Maybe, one day, customers could even pay for haircuts online.

The concept sounds obvious today, but back in 2014, there were many questions and hypotheses that needed testing: Would barbershops take appointments over an app? (The answer initially, for many barbers, was no.) Was the barbershop market large enough to matter? (Yes, despite the challenge of nailing down the exact size.) And perhaps most importantly to investors, were LaRon and Salvant the right founders for this mission?

On the surface, it seems unlikely that LaRon, an attorney, and Salvant, a banker, could pull off a technology startup in the barbershop industry. Neither had experience coding, building products, raising money, or leading teams, never mind cutting hair. In their first attempt to build the app, the founders were nearly scammed out of $5,000 by someone posing as an engineer. Despite their shortcomings, LaRon and Salvant had the most important trait of any startup founder: an experimentation mindset. They approached building Squire like a pair of scientists—creating hypotheses, devising tests, learning, and iterating.

LaRon and Salvant eventually found a legitimate software developer to create the first version of the Squire app, but they struggled to find barbershops willing to test it out. So they set up a barber chair inside their WeWork office on 220 Broadway and brought in barbers to give haircuts. Each appointment became a learning opportunity that helped them refine the app and user experience. It also proved to the founders that there was demand for their product—at least on the consumer side. Around this time they met Jesse Middleton, the co-founder of WeWork Labs and now my partner at Flybridge. LaRon and Salvant impressed Jesse with their grit and passion, and in turn Jesse organized some friends to invest the first $150,000 into Squire. The founders then parlayed this "social proof" into another $150,000 investment a month later.

With the vote of confidence and influx of cash, LaRon and Salvant expanded throughout the NYC barbershop community, but they soon hit another existential roadblock: inertia. "Barbers would use the app if we sent them new business, but they weren't converting their existing customers into Squire users," said Salvant. "They ended up having two systems in place, and the feedback we were getting was, 'I need something to manage my entire barbershop.'"

In order for Squire to succeed, the vision would have to evolve from "the OpenTable for barbershop customers" to an end-to-end finance and management platform for barbershop *owners*. They would compete against Mindbody and StyleSeat for scheduling, Square and Clover (and cash) for point-of-sale, QuickBooks for reporting, ADP for payroll, Yelp for customer discovery, and Mailchimp for CRM. If they could pull it off, Salvant and LaRon estimated their suite of solutions for barbershops would be worth $1,000 per month, per customer. With over 300,000 barbershops in the country—and 10,000 in NYC alone—this new vision presented a multi-billion-dollar opportunity.

This is when Salvant and LaRon invested in their Chelsea Market laboratory. When a barbershop owner—an early

Squire customer—told the founders he wanted to retire, the duo decided to buy out his lease and run the shop themselves. This move cost the company $20,000 at a time when they had less than $40,000 in the bank, but LaRon and Salvant knew the experience could help them perfect their new management platform. "Running that shop was a major learning experience," said LaRon. "We really built the product based on what we learned there."

Over the next three years, the Squire founders would clear every obstacle in front of them with ingenuity and grit. As first-time founders and Black men, they faced more skepticism and outright pushback than their white counterparts. They slogged through multiple rounds of fundraising at less-than-ideal terms, constantly addressing questions about their competence and market opportunity. Despite all of this, the founders secured an $8 million Series A in February 2019 and a $27 million Series B in March 2020, which valued the company at over $85 million. Grinding through years of hard work helped Squire survive and expand through the Covid-19 pandemic.

Today, the company services more than 3,000 barbershops across the United States, Canada, and United Kingdom. In 2023, they processed over $1 billion in transactions and the company was most recently valued at over $750 million.

LaRon and Salvant began their startup journey a few years before the advent of generative AI, but their story illustrates a more fundamental pattern we see across all successful startups: a willingness to experiment. They made small bets each step of the way, doing things that didn't scale (to paraphrase Y Combinator co-founder Paul Graham) in service of learning. LaRon and Salvant's experience was a trial by fire. It's what makes their success so inspiring.

Building a startup is nothing like running a traditional business. To understand the difference, let's start by breaking down the traditional business model.

Deconstructing the Business Model

Startups are businesses without working business models. The mission is to build a working business model—a harmonious mix of business functions and stakeholders—before the startup runs out of money or a competitor beats them to it.

Every business is comprised of the same parts: a customer value proposition, go-to-market (GTM) plan, technology/operations infrastructure, and a monetization strategy (ideally one that eventually makes the business a profit). These four parts together are typically called a business model. Business models need people to operate them (yes, even in the age of AI): stakeholders like the founders, investors, teams, and partners. My HBS colleague, Professor Tom Eisenmann, captured this relationship between business functions and stakeholders in his Diamond-and-Square Framework:

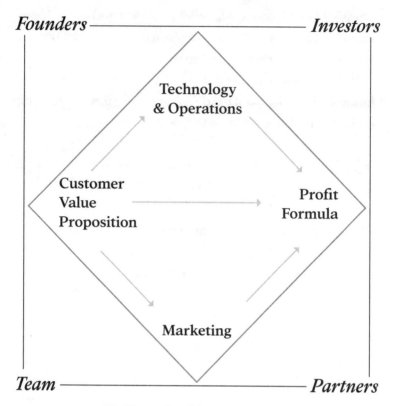

The Diamond-and-Square Framework

These are the essential pieces of every money-making enterprise on Earth, from your favorite corner barbershop to Amazon. In a successful business, these parts are more or less working in harmony. In struggling businesses, one or more of these parts are broken or missing entirely.

Startups are different. They are discovering and creating entire elements of the business model on the fly, often in a messy fashion.

Let's think back to the Squire case study. Founders Songe LaRon and Dave Salvant had an initial Customer Value Proposition—the OpenTable for barbershops—that proved to be incomplete. That didn't make Squire a failure; it made it a startup. Later, the founders struggled to find an efficient GTM strategy. They hoped advertising and consumer-driven word of mouth would grow the business, but LaRon and Salvant eventually settled on outbound sales as their primary GTM motion.

What about technology and operations? I shared how Squire was almost conned out of $5,000 by a fake engineer. The founders also tried hiring numerous technical leads in the early years without success. As non-technical founders, LaRon and Salvant cobbled together barely working MVPs of their app to continue learning and experimenting. In 2017, after finally gaining some traction, the duo knew their app needed an engineering overhaul to keep up with demand. They opted to move to Buffalo, New York, to secure $650,000 in funding from an accelerator and lived there for a year. Like many startups, Squire's early days were defined by broken or insufficient products and technology that didn't scale.

Squire's monetization plan was no less uncertain for the first half decade of the startup's existence. The founders originally planned to charge a $1 transaction fee for every scheduled appointment, just like OpenTable did for restaurant reservations. But when Squire's vision expanded to the backend platform for barbershop owners, LaRon and Salvant switched to a subscription model plus transaction fees. The Covid-19 pandemic forced Squire to revisit its monetization strategy yet again. To support their barbershop clientele, the founders chose to temporarily waive subscription fees, which made up 13 percent of their revenue and the majority of their gross margin at the time. Squire's investors strongly advised against the move, but LaRon and Salvant earned enormous amounts of customer goodwill, which fueled Squire's growth once barbershops opened back up for business.

A startup's human resources are just as precarious as its business components. A successful startup needs the right mix of founder-market-fit, investors, team members, and channel partners. None of these stakeholders are guaranteed in the early days.

So how on earth does any startup get off the ground? It's a race against time and money—often not even enough to support the founders full time. Anyone approaching building and running a startup as they would a normal business is doomed to fail. Startup founders need a different operating model to turn their ideas into bona fide businesses. I call this model the Experimentation Machine.

The Experimentation Machine: An Operating Model for Startups

A startup founder's job is to build an experiment-driven organization, i.e., an Experimentation Machine.

The goal is to test and probe for answers to the most fundamental questions of any business: Who is my ideal customer? What do they want and need? How do I deliver it to them? How do I find more ideal customers? How do I make money and build a sustainable business?

The Science of Startups: How to Build an Experimentation Machine

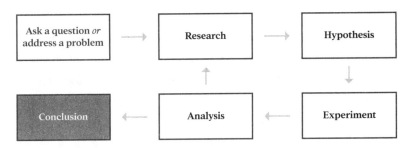

Following the Scientific Method

Building an Experimentation Machine requires a basic understanding of the scientific method. It starts with asking a question or identifying a problem: *Why is it so hard to get a haircut?* Next is research: How are things done today? What are the alternatives and workarounds to this problem? What do people actually want? When LaRon and Salvant first had their idea for Squire, they started by surveying as many barbers and barbershop customers as they could find. This informed their earliest product ideas.

After conducting deep customer discovery—the process of thoroughly researching potential customers' needs and problems—the founding team creates a hypothesis or an educated guess. Would barbershop customers book appointments through an app? Would barbershop owners adopt a new platform for scheduling? Founders need to make hypotheses for every part of their business: customer value proposition, go-to-market, technology infrastructure, and monetization plan. A hypothesis must be falsifiable. Can you definitively prove it to be correct or incorrect?

Then you identify the most important hypothesis and design an experiment to test it. LaRon and Salvant created a basic mobile app and set up a barber chair inside WeWork. They determined that customers wanted an app for scheduling appointments, but barbershop owners were reluctant to adopt another tool (especially one that cost money). The founders then built a lab to run even more experiments—the Chelsea Market barbershop—and learned what was needed in a management platform for shop owners.

Finally, founding teams must analyze and implement the results of the experiment. There are no "failed" experiments. The only goal is to learn, therefore the only way to fail is to not take anything away from a test. Then the process starts again: question, research, hypothesize, experiment, and analyze.

Building a startup is like navigating an uncharted territory. In my last book, I dubbed this territory *Startupland*. The journey through Startupland goes through three distinct phases:

- *The Jungle:* The land is wild and dangerous. There are no paths and no set direction. All you have is a machete to cut through the brush.

- *The Dirt Road:* After trial and error, you eventually clear a way through the jungle. The road is rough and winding, but at least you know generally which way to go.

- *The Highway:* Through sheer effort and a little bit of luck, you pave your dirt road into a highway, where you can move fast and get to your final destination (relatively) safely.

Startupland has no map to show you the way, but you do have a compass: the Experimentation Machine framework. By testing and iterating, you will eventually find the answers to build a functioning business model. The challenge is, can you do it quickly enough before the jungle swallows you up?

Test Selection as Strategy

The Experimentation Machine is the basic operating model for startups, but what tests should you run and when? In an environment where you have limited time and even more limited resources, test selection becomes the most important strategic decision.

Startup founders should not expect to discover a working business model right away. Rather than try to carve a highway out of a jungle, they should think about their progress in terms of key milestones: Which experiments will unlock the founder's ability to raise more capital and attract better resources?

The strategy of test selection boils down to three essential questions:

1. Which business model component is **most controversial**, and what is the essential hypothesis for that component?

2. What is the **key milestone** I need to achieve that will lead to my next valuation inflection point, helping unlock more capital from investors?

3. Where does the *greatest risk* exist in my business model, and what does the flow of dependencies look like?

The answers depend on the startup and their stage of growth. However, when starting from scratch or running the Experimentation Machine framework for the first time, startups should run tests in the following sequence:

1. *Customer Value Proposition experiments:* Who are your ideal customers, what do they want, and how does the startup deliver? The primary business function under scrutiny is the product.

2. *Go-to-market experiments:* What is the ideal sales model? What are the most efficient customer acquisition channels? What is your best growth flywheel? Product, sales, and growth are the key business functions to test in this stage.

3. *Business model/profit formula experiments:* How do you monetize and price? How do you improve your margins? Are you maximizing your potential valuation? Marketing, customer success, and business development now enter the picture.

This sequence answers the most fundamental questions about your startup in the order needed to build a functioning business. Your monetization strategy does not matter if you can't attract customers, and your go-to-market strategy is useless without having a strong customer value proposition. We will dive into each of these experimental sets in Chapters 4 through 6.

Finally, startup founders must check and re-check their Experimentation Machine often to ensure it's in working order. It is up to the leadership to determine that experiment sequencing, design, and execution are running optimally. Here are a few important questions to ask yourself:

Sequencing →	**Design** →	**Execution**
Is this experiment the right one to focus on at this exact moment?	Have I narrowed the experiment to focus only on the Most Important Thing at this exact moment?	Am I set up to execute this experiment properly?
Are there dependencies I need to think through?	Is the experiment tightly defined and straightforward?	Do I have the right resources and team?
	Will the experiment yield a falsifiable outcome?	Are we instrumented to collect the necessary data to evaluate the results?

A startup founder's job is that of Chief Experimentation Officer. You must guide your team to adopt the scientific method and run the Experimentation Machine operating model diligently. When done properly, like Songe LaRon and

Dave Salvant, you will have a chance to find your way out of the startup jungle and onto a dirt road. With a bit more luck and a lot of hard work, you may pave a highway that leads your startup to scale and financial exit.

From a thirty-thousand-foot view, the journey looks like this:

	#1 Value Prop Fit	#2 GTM Fit	#3 Biz Model Fit
Goal of Phase	Customer Love (40% Test)	Repeatable Acquisition	Unit Economics, Growth & Strategic Moat
Target Market	Innovators	Early Adopters	Early Majority
GTM Process	Founder Selling	Playbook Discovery	Partners/Sales Team
Sales Leader	Founder	Player-coach	Process/Team Builder
Demand Gen.	Personal Network + Referrals	Paid Mktg Experiments, Partner/Channel Experiments	Multi-Channel Partner with Sales
Pricing	New: Price Low, Prove Demand / Existing: Price High, Prove Willingness to Pay	Solve for Breakeven	Solve for Profitable Unit Economics
Risks	PMF False Positives, "Tech Crunch" False Positive	Vanity Metrics/False Positives, Runway to Pivot	Premature Scaling

Putting It All Together

I share this so you have a sense of the big picture as you dive into the book. The details of these phases will be discussed in the coming chapters.

Each of these stages—CVP fit, GTM fit, and Biz Model fit—is a key milestone in the startup journey. They are chances to raise funding and attract talent. As a VC, I evaluate startup investments with these benchmarks in mind and look for evidence of future valuation inflection points. Everyone in my field does. Founders should keep these stages in mind as well when making strategic decisions. Don't invest in a sales team before you have found a working CVP. Be wary of scaling prematurely even after you find a successful business model.

The Experimentation Machine model will not guarantee startup success, but it gives you the best opportunity. In a deeply uncertain environment, the best strategy is to place small bets and iterate along the way.

Moving Faster with AI

Speed is critical to a startup. The more experiments you can run, the more likely you are to discover a functioning business model. For this reason, generative AI represents the greatest leverage-booster for startups since the Internet. We are entering a new era of startups, where founders are accomplishing much more with much less, rapidly accelerating the search for product-market fit. Let me now introduce you to the 10x Founder.

Chapter 2

The 10x Founder: Harnessing AI to Create Startup Superpowers

I N FEBRUARY 2024, OPENAI co-founder and CEO Sam Altman made this prediction during an interview at a JP Morgan conference:[3]

"We're going to see ten-person companies with billion-dollar valuations pretty soon ... In my little group chat with my tech CEO friends, there's this betting pool for the first year there is a one-person billion-dollar company, which would've been unimaginable without AI. And now [it] will happen."

Startups are defined by their constraints: They have limited time, money, talent, resources, and opportunity compared to established businesses. But their main advantage—their ace in the hole—is their ability to move fast. As we covered in Chapter 1, speed is imperative as startups search for a functioning business model. And now with the power of generative AI, founders can—and will be expected to—move even faster. Welcome to the era of the 10x Founder.

The startup world has long mythologized the 10x developer: The computer engineer who is so good that they can do the work of 10 average engineers. 10x developers were once the difference between a startup becoming a unicorn and fizzling out.

10x Founders have their own set of superpowers: They have mastered the relatively new skill of AI delegation to build faster, leaner, and more profitable startups than ever before possible. Using AI, they run experiments more quickly to out-learn and out-iterate their competitors. They scale their companies with one-tenth the headcount (or less). In short, 10x Founders use AI in creative ways to improve every function of their business.

The best way to understand a 10x Founder is to introduce you to a few of them. At Flybridge's Founders Week 2024 Conference, we featured numerous 10x Founders who are using AI to grow rapidly and efficiently. Here are the stories of two startups that presented at the conference:

Unlocking a Fragmented Market: Nick Ornitz and Shannon Kay, Topline Pro

One of the major themes of AI is how it will unlock new markets and business opportunities. Topline Pro, an all-in-one website builder and marketing provider for home services companies, would have been nearly impossible to build before generative AI. Co-founders Nick Ornitz and Shannon Kay started their company in 2020 as Dwelling, a video chat service for plumbers and homeowners. They even dropped out of HBS (which, for the record, I did not encourage!) and entered Y Combinator to build their vision. But when plumbers said their biggest pain was finding new business, Ornitz and Kay pivoted. They rebranded as Topline Pro and re-launched in January 2022, using GPT-3 (the OpenAI model *before* ChatGPT) to rapidly create SEO-optimized websites for home service providers.

Today, Topline Pro's mission is to serve the highly fragmented market of over five million service-based businesses in the United States, including plumbers, electricians, general contractors, landscapers, and more. As of July 2023, Topline Pro had helped generate $180 million in new business for thousands of home service providers around the country. They are doing this with a team of less than forty humans.

With a highly fragmented, hyper-local market like home services, the biggest challenge is reaching your target customers in an efficient manner. The "marketing for home services" market is also highly fragmented, dominated by local freelancers and agencies. But Topline Pro has created an AI-powered outbound sales strategy that connects them to thousands of new small businesses a month and has powered their growth.

First, Topline Pro uses ChatGPT to create "SparkNotes" of every small business in a local market. They scour Facebook groups like "Lawncare Mafia" and score every profile to identify the most qualified leads. "It starts with how we qualify and find leads," said Ornitz. "We look at all the different profiles of home service businesses online and then we feed the history of those business profiles through an internal tool that uses GPT to identify themes about the business." This information is added to the company's CRM and used by Topline Pro's salesforce.

Then Topline Pro creates personalized sales content using their GPT-powered tools. It's hard to get in touch with home service business owners because they rarely check their email, but Ornitz and Kay thought they could crack the code. After early attempts at cold email outreach—in which they achieved a whopping *zero* percent response rate—the startup tried adding personalized, AI-generated messages. Their response rate jumped to 10 percent across their best email campaigns. Topline Pro now sends out a few thousand AI-generated messages per week, *per sales rep*. Once they receive a response from a prospect, the sales rep takes over the conversation.

Finally, Topline Pro uses AI to automate the customer experience. They use an AI-powered customer support chatbot that can answer their customer's questions 24/7. Topline Pro customers can interact with this chatbot via text. As of this writing, over 75,000 text messages have been exchanged between customers and chatbot without human intervention.

Ornitz and Kay are exemplary 10x Founders. They didn't let the outdated beliefs stop them going after a massive, fragmented market. They were right on the cutting edge of using generative AI and continue to improve their business functions with each new feature release. Instead of employing hundreds of people, they are conquering a multi-billion-dollar industry with just a few dozen creative, versatile employees.

Scaling Founder-Led Support: Matt Mireles, TalkTastic

As a self-taught, "pseudo-technical" founder with a background in AI, Matt Mireles had a 10x Founder mindset long before the advent of ChatGPT. Now he has even more leverage to build exceptional customer experiences at scale.

Mireles' startup, TalkTastic, is developing a voice-first interface for computers. Ninety-six hours before launch, Mireles realized with dread that he had no customer support workflow in place. He didn't just want to tack on a traditional chat-based system; that type of customer experience was the exact thing he was trying to eliminate with TalkTastic. Mireles wanted to create a voice-first support system that was available 24/7, without hiring a team of support reps.

His elegant solution is worthy of an HBS case study on its own. When customers click on the support chat button, they are greeted by a friendly video from Mireles, who explains TalkTastic's recent updates and how to use their customer support system. Then customers are prompted to either leave a voice message or type out their question in a chat box using Typeform's VideoAsk tool.

Once a message is created, the voice file is fed into ChatGPT to be transcribed, then added to Intercom (which Mireles uses as their support system of record). Then the message is fed into Anthropic's Claude AI, which writes out a personalized response and feeds it back into Intercom. In about five minutes, a customer receives a personalized message back, all without human intervention. ChatGPT also analyzes the customer message to determine if the issue is a bug. If it is, a new ticket is created in Linear (TalkTastic's project management tool) for triaging. This entire system is connected using the workflow automation tool Zapier.

Mireles built this video-first, AI-powered support workflow in just four hours, without writing a single line of code. It's not just a placeholder, either. Customers regularly praise Talk-Tastic's support experience, with one user saying, "The UI for feedback is stellar and is deserving of a product of its own."

TalkTastic's founder-led support model would have been impossible before the dawn of generative AI. A founder's time is precious, and as much as Mireles wants to help customers, he has other problems to address. He sees this AI-powered workflow as a step toward a future where founder-level, personalized support can be provided at scale, potentially even using AI avatars. "When I look at the brands today, they're increasingly about personalities. They're about people," Mireles explains. "You talk to the CEO, you feel like, 'Oh wow, okay, there's a human being here. I have a relationship,' and I believe it can lead to deeper loyalty and retention."

The startup world is quickly being divided into Old World and New World thinking. Old World thinking says that Mireles has to settle for a chat-based support system because he doesn't have the time or resources to build something better. New World thinking—10x Founder thinking—says, "give me four hours and an LLM." Old World thinking says you need hundreds of salespeople, organized by region, to attack a fragmented industry. New World thinking says, "I can send personalized videos to thousands of prospects per week in just minutes."

I share these stories because it's hard to break out of the Old World mindset, even when you come into contact with the power of AI. We need to see 10x Founder thinking in action to fully understand the leverage we now have at our fingertips. These stories should inspire you to think bigger about using AI in your business.

Here is one more story to inspire you, this one from a friend's startup still in stealth mode. This founder recently lost his technical co-founder. In the past, that would have been an existential crisis. Today, though, my friend has opted to use ChatGPT to build his MVP. As he shared with me, "The result is that my burn rate is incredibly low, and velocity—the speed at which I can build and iterate on my product—has shot through the roof." He has been able to test out his initial hypotheses while searching for a new technical co-founder on the side.

Economists will try to measure the definitive impact of AI on entrepreneurs and economic productivity for years to come, but the early evidence is encouraging. In one study led by my HBS colleague Professor Rembrand Koning, high performing entrepreneurs in Kenya were able to improve their business performance by 15 percent simply from AI *advice*. As the models get much better, that advice—and the reasoning behind it—will have an even greater impact.

Traits of a 10x Founder

Let's try to deconstruct the anatomy of a 10x Founder. What makes Nick Ornitz, Shannon Kay, Matt Mireles, and my stealth founder friend different? There are several traits they embody. I encourage every founder to emulate and develop them.

"Scaling Without Growing" Mindset

When I was a founder, I was obsessed with hiring. As soon as we secured more capital, we hired more people. If we closed another strategic partner, we hired. If we had a new product idea, we hired.

I've been an executive at successful B2B and B2C startups that scaled. I've been an investor and board member for hundreds of startups. In all cases, when things started to work and we hit product-market fit, we scaled rapidly because we hired rapidly. And we hired rapidly because we scaled rapidly. That era is over.

Going forward, founders will be more focused on deploying the right AI workflow than hiring teams of people to execute. Adobe executive Scott Belsky coined a new phrase and I like it. He wrote, "We are entering an era of *scaling without growing* . . . Every function of an organization will be refactored in ways that allow small teams to scale their reach and ambition without growing headcount proportionately."[4] Nick Ornitz and Shannon Kay clearly embody this scale-without-growth mindset. They are tackling a fragmented market at scale with under forty employees. Instead of hiring hundreds of sales development reps (SDRs) to perform cold outreach, they automate the sending of thousands of video messages per week using AI. Matt Mireles of TalkTastic is clearly cut from this cloth as well. He has scaled *himself* to deliver personalized, founder-led support to his customers. And while my founder friend isn't at scale yet, he chose not to hire an expensive technical co-founder in favor of building an MVP himself, using code generated by ChatGPT.

Modern AI tools, and particularly Agentic AI, have upended the traditional headcount needs in virtually every industry. 10x Founders must have strong strategic and managerial skills to deftly orchestrate their AI assets—and key human teammates—to scale rapidly without the massive employment overhead. (We'll dive deep into scaling product with AI co-pilots and agents in Chapter 7.)

AI-Forward and Obsessed with "Better"

At Flybridge, we invest in *AI-forward* startups. That doesn't necessarily mean AI is a core part of the product, but rather the organization is imbued with AI workflows to grow faster, leaner, and more profitably. 10x Founders maniacally re-evaluate and re-invent every part of their business as new capabilities come online. It requires an experimentation mindset to try new tools, rapidly adopt those that make an impact, discard those that don't, and change your SOPs (Standard Operating Procedures) accordingly.

The bulk of this book is focused on refactoring each piece of your business model with AI tools and workflows: customer value proposition, go-to-market, tech & ops, and monetization. There are already dozens (or by the time you read this, hundreds) of AI tools in each of these categories. A friend of mine, the founder of a public software company in the sales and marketing space, tells me he's tracking forty-four companies that are building AI-powered SDRs. Building an AI-forward startup means you must be constantly on the lookout for new tools that will give you an edge.

As a former product manager, one tool that I'm particularly excited about is ChatPRD, created by an experienced PM named Claire Vo. ChatPRD is an on-demand Chief Product Officer that writes and improves product requirements documents (PRDs). PRDs are critical documents to convey the scope of a product to an engineering team, outlining the rationale for building as well as user experience goals and success metrics. It is often a laborious effort for founders and product leaders to write—thus, leveraging AI to save them time is a godsend.

Not all AI tools are going to survive, naturally. But if you wait until the dust settles, you will be leagues behind the 10x Founders who dove in head first and sorted through the slop themselves. Being AI-forward means being on the cutting edge, even when the future is unclear.

Focused on Profit and Efficiency

As a venture capitalist, I can't help but think of the financial implications of generative AI tools in the hands of 10x Founders. Next time a founder tells me they are going to spend their investment money on a team of support agents, I'm going to refer them to Matt Mireles and what he was able to accomplish in under four hours. If they can't comprehend how to use such a powerful form of leverage, they may not be long for the world of VC-backed startups. As an investor, it would be negligent for me not to back founders like Mireles who can get so much done in a fraction of the time and cost. Money will follow the founders who take full advantage of AI, and results will be close behind.

Generative AI is making software companies more profitable while requiring less capital. Some argue that these profits will disappear with more intense competition. In other words, the headwinds of competition and the lack of competitive moat in an AI age outweigh the tailwinds of operational efficiency. Those headwinds are real, but I'm not sure they're strong enough to slow down this productivity boom.

For a glimpse into the future, we can examine the most AI-forward organizations on the planet. At Flybridge, we created an "AI Index" to track the thirty public companies that stand to benefit the most from AI, focusing attention on the most innovative companies in the sector. In the years 2020-2024, a period of nascent AI tools adoption by these pioneering companies, the thirty companies saw massive productivity gains with median revenue per employee growth of 20%. We can see the specific numbers for three of these

AI leaders—Alphabet (Google's parent company), Microsoft, and NVIDIA—in the chart below.

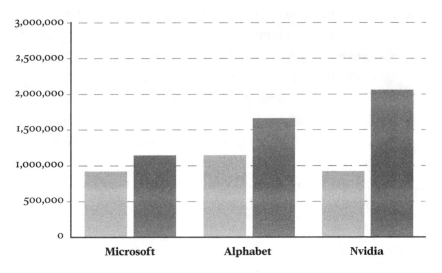

Y: Revenue per Employee ($) by X: Company for ■ 2019 and ■ 2024

I would be shocked if we didn't see this positive effect on productivity ripple through other large organizations (finally defying the confounding Productivity Paradox of IT[5]). Klarna recently reported tremendous efficiency gains in using AI for customer service, with an AI assistant handling two-thirds of their chats. ServiceNow has mandated that every department provide an AI roadmap. The company's CEO is obsessed with using AI for process automation. He declared, "Any process that exists in the enterprise today will be reengineered—or engineered, depending on how messy the process is—with generative AI... Every workflow in every enterprise will be rethought."[6] Amazon CEO Andy Jassy shared in a LinkedIn post that after their generative AI assistant, Amazon Q, was integrated into their internal systems, it saved the company 4,500 developer-years of work (Jassy noted, "Yes, that number is crazy, but real.") and $260 million in annualized efficiency gains. Alphabet CEO Sundar Pinchai recently reported that 25 percent of all new software code created by the company is being created by AI.

These productivity gains are so dramatic, and happened so quickly, that I believe these companies may have reached the point of peak employment. Going forward, all their revenue growth will be achieved with the same number of employees— or even fewer. We will have to wait and see.

No matter how innovative and determined, big companies are slower to adopt new processes than small companies. Culture and organizations are too entrenched and large groups of humans adapt slowly. Startups get to begin with a blank slate. 10x Founders are going to take that blank slate and make magic.

Balancing AI Efficiency with Real Risks

If it's not obvious by now, I am a techno-optimist. Gen AI offers immense potential for startups to accelerate their experimentation and innovation processes, yet it is important to acknowledge the risks and challenges associated with this powerful technology. Apart from some of the ethical implications (which we will cover in Chapter 10), this book won't explore many of these risks in depth; that is a complex and evolving topic that merits dedicated exploration. However, you should be aware of them at a high level and consider how you can use these tools thoughtfully and responsibly.

Fighting Human Bias in AI Models

One of the primary concerns surrounding AI is the potential for human bias to permeate the models. As these models are trained on vast amounts of data from the Internet and other sources, they can inadvertently learn and perpetuate racist, sexist, bigoted, and hateful behavior we see online. This selective training set can lead to the proliferation of content and analyses that reinforce stereotypes or discriminate against certain groups. Startups must be aware of these risks and take steps to avoid propagating bias in their AI use.

There is a famous case of Amazon shutting down an AI-based recruiting tool a few years ago when they realized that it discriminated against female candidates. Their training data set? Résumés submitted to the company over a ten-year period. You can easily guess what gender the majority of those résumés represented, thereby creating a system that had an inherent gender bias due to the training data.

Biases in training data underscore the importance of transparency and *explainability*—the ability for an AI model to articulate how it developed an output. One of my portfolio companies, Zest AI, is a leader in AI-driven credit underwriting software. You can imagine how important it is for their customers—banks and credit unions—to have clear transparency and explainability on loan decisions that the model recommends to ensure fairness and regulatory compliance.

Privacy, Data Protection, and Copyright Infringement

AI models also raise concerns about privacy, data protection, and copyright infringement. As these models require vast amounts of data for training, there is a risk that sensitive or personal information could be inadvertently included or exposed. There is also the very real risk that copyrighted material will be co-opted without permission. *The New York Times* refers to the way AI companies have trained their models on copyrighted works as "the original sin of AI"[7] and has joined seven newspapers in suing OpenAI and Microsoft. Major record labels have also sued AI music generators. Startups must ensure robust data governance, security practices, and plagiarism checks to protect user privacy and comply with relevant regulations and copyright laws.

Malicious Applications and Security Breaches

We may not be wiped out by an AI supercomputer with a vendetta, but these tools in the hands of bad actors is a serious concern. Malicious users are already leveraging AI to generate fake news, deepfakes (computer-generated media that appears to be a real image or video), and other deceptive content, which can have severe consequences for individuals, organizations, and society. Furthermore, AI outputs are not perfect, and using AI for tasks such as code generation or business automation could introduce security vulnerabilities and other unintended consequences if not validated and monitored (we also have a portfolio company for that!). Weaponizing AI has become an area of concern—not only from individual actors but also nation-states. Many are calling for more regulation and transparency of AI models. Those calls led the European Union to pass the first-ever legal framework to address the risks of AI in 2024.[8]

Environmental Impact

Finally, there's the environmental impact of training and deploying large-scale AI models. The computational resources required for these models demand significant energy consumption, which contributes to carbon emissions and climate change. A Carnegie Mellon University study estimated that generating one AI image takes as much energy as fully charging your smartphone.[9] The International Energy Agency projects that data centers' electricity consumption in 2026 will be double that of 2022.[10] As AI scales, there needs to be efforts to make its environmental footprint more sustainable.

The good news is that the most senior leaders of large AI companies, as well as senior government officials, are all very aware of these risks. From my conversations with many of them, I am confident these problems are being worked on assiduously. Further, the rise of open-source AI models can help mitigate some of the risks. Many leaders, like Meta's Mark Zuckerberg, have advocated for open-source models to provide greater transparency and avoid getting locked into one company or another's closed ecosystem.

One issue that we worry about at Harvard is whether our students will get complacent and use AI to answer questions and solve problems without digging deeper. In one of my classroom conversations, I challenged a student on her answer to a question I posed and what supported her point of view. Her response? "That's what ChatGPT told me!" We can't allow AI to fully take the place of critical thinking and first principles understanding of issues and topics.

The pace of progress and improvement of these systems is hard to comprehend. Humans are strong linear thinkers. We are not great at exponential thinking. At a recent talk at Harvard, Sam Altman observed that it is far too easy to underestimate these models, and it will shock people "just how good these models are becoming" in the very near future. As of this writing, ChatGPT4o is the current state of the art (along with competing offerings from other software providers like Llama-3.3, Gemini 2.0, and Claude 3.5). By the time you read this, ChatGPT5 or even ChatGPT6 might be widely and freely available. These tools are simply getting better and better and better. One former OpenAI researcher forecasted in a June 2024 essay that based on the trendlines in computational power, algorithmic efficiency, and other advancements, super intelligence would be achieved by 2027.[11] Microsoft's top AI executive, Mustafa Suleyman, writes in his book *The Coming Wave*, "It's not just a tool or a platform but a transformative meta-technology, the technology behind technology and everything else, itself a maker of tools and platforms, not just a system but a generator of systems of any and all kinds."

Again, entrepreneurs would be well served to learn to harness these powerful systems now—recognizing that they will only get more and more powerful with time.

The Search for Product-Market Fit

10x Founders will use generative AI to rapidly experiment, iterate, and scale their startups. But executing quickly doesn't matter if you're aiming in the wrong direction. We need to talk about product-market fit, that crucial milestone on every startup's journey to success.

But is it a milestone? Or is it a more like a HUNCH?

Chapter 3

Having a HUNCH: Redefining Product-Market Fit

EVERY THANKSGIVING, Harvard Business School gives away free pies to faculty and staff. It's kind of a big deal. Eager administrators and professors, myself included, form lines out the door and down the sidewalk in front of the Chao Center on campus. Quantities are limited, so you risk missing out if you aren't timely (and then you might as well not show up to Thanksgiving dinner). Given it's late November in Boston, the temperature is anywhere from cold to *really* cold with a mix of wind, rain, sleet, or snow. Regardless, people always line up and wait for their pies.

Now I have a question for you: Does HBS have product-market fit (PMF) with their pies? The most common definition of PMF, from venture capitalist Marc Andreessen when he coined the phrase in 2007, leaves plenty of room for interpretation:

> "Product/market fit means being in a good market with a product that can satisfy that market."

Now, obviously a market comprised solely of HBS faculty and staff is a little small for a venture-backed startup. But the pies clearly satisfy the market. If they didn't, people wouldn't line up around the corner in terrible weather to get one. Still, this "business model" is missing some key components, such as revenue. Yet giving away your product for free is such a common strategy in Startupland that it's become a joke (Watch the episode "Bad Money" from the HBO show *Silicon Valley* for my favorite rendition of this schtick).

We will discuss the implications of "freemium" in Chapter 6: The 10x Club. For now, let's keep our focus on defining product-market fit: what is it, really, and how do you know when you have it?

Let's explore a slightly more relevant example of a startup in search of PMF to see if we can start to answer these questions. Who's ready to dance?

ClassPass: When Product-Market Fit Nearly Kills Your Business

Just like Squire, ClassPass was born out of frustration.[12]

For Payal Kadakia, it was the frustration of booking a fitness or dance class. Kadakia, the daughter of Indian immigrants, had a lifelong passion for dance, one she continued to pursue while attending MIT for mathematics and operations management. She even founded an Indian fusion dance troupe as an undergrad. After graduating, she spent a stint in consulting at Bain & Company before progressing to a strategy and business development role at Warner Music Group in New York City.

It was during this period that Kadakia noticed how difficult it was to find and book dance and fitness classes. One night in 2010, after a fruitless search for a class, Kadakia hit her breaking point. "An hour later, I was so tired of searching, checking schedules, mapping, scrolling, and clicking to find the right class, I didn't even end up going to a class—simply out of frustration," said Kadakia. "I realized the problem: why isn't finding a class as easy as booking a reservation on OpenTable, or finding shoes on Zappos?" (OpenTable was a popular source of inspiration in the 2010s.)

Kadakia reached out to a childhood friend and fellow dancer, Sanjiv Sanghavi, who had also experienced the pain of booking classes in NYC. He decided to leave his Wall Street job to join Kadakia on her quest. In 2012, they launched Classtivity, a fitness studio search engine and online booking platform. Their promise to customers was compelling: finding a fitness class shouldn't be more difficult than taking one.

While the founders identified a painful problem, they struggled to put together the right solution. The search engine concept was a bust. As Sanghavi recalls, "People were searching for classes, but they weren't doing the thing we needed them to do to make any money—booking. We weren't really making any revenue."

(Classtivity had a popular product that couldn't be monetized. Sound familiar? It's safe to say that neither HBS's pie team, nor Classtivity at this stage, had product-market fit.) Abandoning the search engine model, the team decided to emulate another red-hot startup of the day: Groupon. The coupon-based e-commerce company was one of the fastest-growing startups in history. If the concept worked broadly for small businesses and retailers, why wouldn't a specialized platform for fitness and dance classes be a hit? Classtivity created bundles of fitness "tickets" at various studios for a one-time flat fee. Studios opted in to give away these first classes for free with the hope that guests would convert into paying members.

The coupon-based model took off. By April 2013, Classtivity was hosting 15,000 listings in eight cities and had generated $50,000 in revenue. "We got more engagement with [the coupon bundle] in the first three or four days than we had with [the search engine] in six months," Sanghavi said.

But Classtivity's coupon model had the same issues that eventually toppled Groupon: Visitors did not convert into regular customers. Many of the coupon buyers were what Kadakia called "dabblers." "Our vision was to get you habituated to going to the same studio over and over again, but our customers were treating it like a one-month experience." For fitness studios, the value proposition quickly lost its luster. They were giving away free classes while Classtivity captured all the value. Many pulled out of the program. A two-sided marketplace does not have product-market fit if one side is not satisfied.

Classgoers, however, loved the product, and some even wanted to buy multi-month packages. This sparked yet another idea for Kadakia and Sanghavi: a subscription-based model that offered unlimited classes and shared revenue with studio owners.

Kadakia was hesitant to pivot yet again, but the coupon-based model was unsustainable. After some difficult conversations with her team (including, as she shared with me, an existential weekend executive team meeting), they decided to make the switch to monthly subscriptions. They also rebranded the company as ClassPass.

That's when things really took off. The company grew to 500 studio owners and thousands of users who generated 500,000 reservations in just a few months.[13] Classgoers loved the simple subscription model and studio owners were making their fair share.

So was this the moment that ClassPass found product-market fit? They had found a good market and created a product that satisfied *both* sides of the market, and they grew rapidly as a result.

Not quite. ClassPass ran into another existential problem: Their most active customers were killing the company. ClassPass collected a fixed monthly fee from users, but their costs were variable. Every time a user took a class, ClassPass dutifully paid the gym its fee. A small percentage of power users took dozens of classes per month and were costing the company money. The more classes that users attended, the worse the unit economics became.

Ironically, fitness studios faced the same issue: Their most active users cost them the most money. All the profit is made on users who don't attend classes, so gyms oversell subscriptions and pack their classes to discourage people from attending. This business model was anathema to Kadakia's original vision. As she wrote in a blog post to her community in 2017: "The impact on our business was unsustainable . . . we'd be sabotaging the vision at the very heart of this company."

So the startup made yet another change to their business model by introducing price tiers and getting rid of unlimited classes. Now, everyone was happy: variety-seeking fitness enthusiasts were still signing up and happy to pay a bit more, studios now had regular customers, and ClassPass no longer had to worry about growth negatively affecting their profitability. The company went on to grow substantially, raise hundreds of millions of dollars, and was acquired by Mindbody for a reported $1 billion in 2021. The combined Mindbody ClassPass is expected to reach $500 million in revenue in 2024.[14]

Levels of Product-Market Fit

The ClassPass case study shows the limitations in the way we traditionally define product-market fit—namely, it's defined too narrowly and viewed as a one-time milestone. We talk about startups as pre-PMF and post-PMF, but the reality is that PMF is an ongoing process. All companies are in search of *stronger* product-market fit at all times. ClassPass (and Classtivity before it) showed signs of PMF at every stage, but the founders continued to experiment to find a better business model, eventually landing on a non-obvious solution that became extremely successful.

We need a better definition of product-market fit that is more comprehensive and more quantifiable. Why? Because the only thing more dangerous than not finding PMF is *thinking* you have PMF and scaling prematurely. If Kadakia and her team had chosen to pursue the coupon model (after all, it was selling well), it's hard to say if they would have survived. Time is your most valuable resource; you can't waste it chasing dead-end ideas.

First Round Capital created a four-stage definition of PMF that better fits the more dynamic reality for startups:[15]

1. *Nascent*: You have a handful of somewhat engaged and happy initial customers, but things still feel early and messy.

2. *Developing*: You have more engaged, paying customers and less churn—you need to work on driving demand.

3. *Strong*: Momentum is picking up and you're finally feeling the "pull" of demand. It's time to focus on increasing efficiency.

4. *Extreme*: You're repeatedly and efficiently solving an urgent problem for a large number of customers who need your product.

ClassPass went through each of these four stages of PMF:

1. Nascent: Classtivity search engine
2. Developing: Classtivity coupons
3. Strong: ClassPass Unlimited Subscription
4. Extreme: ClassPass subscription tiers

Calling yourself post-PMF can lull you into contentment and put you at risk of optimizing for the wrong business model. Recognize that you can always increase the strength of your product-market fit and should continue experimenting.

But these levels only get us halfway there to a better definition of PMF. We also need to evaluate our startups across a broad set of metrics. In other words, we need a HUNCH.

HUNCH Metrics for Defining Product-Market Fit

Many startup founders and investors use customer demand as a simulacrum of product-market fit. But HBS's pie giveaway and ClassPass's unlimited subscription model prove that demand is only part of the equation. In my classroom and with our portfolio companies, I define product-market fit as having a HUNCH.

HUNCH is an acronym for five key metrics that comprehensively test for product-market fit:

- **H**air-on-Fire Customer Value Proposition (as defined by the 40 percent test)
- **U**sage High
- **N**et Promoter Score
- **C**hurn Low
- **H**igh LTV:CAC Ratio

There are benchmarks for each of these metrics to help you define which level of product-market fit you've achieved, from Nascent to Extreme. I'll share those benchmarks at the end of the chapter. For now, let's talk more about each of the HUNCH metrics:

H—*Hair–on–Fire: Customer Value Proposition (The 40 Percent Test)*

How badly does your customer need your product? The first priority in finding PMF is uncovering a problem so urgent, it's like your customer's hair is on fire.

Michael Seibel of Y Combinator likes to jokingly extend the metaphor: "If your friend was standing next to you and their hair was on fire . . . it wouldn't matter if they were hungry, just suffered a bad breakup, or were running late to a meeting . . . they'd prioritize putting that fire out . . . If you handed them a brick, they would try to hit themselves on the head to put out the fire. You need to find problems so dire that users are willing to try half-baked, V1, imperfect solutions."

> *A Customer Value Proposition (CVP) is your unique solution to a painful problem for a specific target customer. It's WHO buys from you, WHAT they buy, and HOW you deliver it.*

We'll cover exactly how to experiment your way to a strong CVP in Chapter 4. For now, let's discuss it as a key indicator of product-market fit. How do you quantify the strength of your CVP? Before you launch your product, all you have are hypotheses. But customer discovery and research-driven ideation can help.

Here is a question I encourage founders to ask target customers during customer discovery: "If this product existed, would it become a top-two priority for you to evaluate?" It's not just about whether or not they would try the product. Given time and budget constraints, what would the user stop doing—or stop purchasing—in order to free up time and budget to try your product or service? No one has time for priorities number three and four!

But you can only learn so much from customer interviews. This is why so many smart people recommend creating a minimal viable product, or MVP; the faster you get a product into customers' hands, the faster you will learn what is working (and not working). Measuring the strength of your CVP becomes much more quantifiable once you have a product in the world. Sean Ellis, an entrepreneur and growth marketer, created a brilliantly elegant experiment to test this. It's simply called the 40 Percent Test. It's a one-question survey to existing users:

> *How would you feel if you could no longer use this product?*
> **a.** *Very disappointed*
> **b.** *Somewhat disappointed*
> **c.** *Not disappointed*
> **d.** *I no longer use this product*

Then you measure the percentage of users who answered "very disappointed." After benchmarking over 100 startups, Ellis determined that 40 percent is the magic number that determines a strong CVP (I will put this number in the context of PMF levels later in the chapter).

Superhuman, the email productivity app, successfully used the 40 Percent Test to iterate on their early product.[16] When founder Rahul Vohra first surveyed their users in early 2017, only 22 percent answered "very disappointed." But when Vohra looked into the data, he noticed something interesting: The users who answered "very disappointed" were mostly startup founders, executives, business development professionals, and managers. These personas were Superhuman's best customers—businesspeople who "lived" in their email inboxes. When filtering the data by these personas, 33 percent answered "very disappointed" to the question—a massive leap. Vohra and his team narrowed their focus to these target customers. They asked three more follow-up questions:

> - What type of people do you think would most benefit from Superhuman?
> - What is the main benefit you receive from Superhuman?
> - How can we improve Superhuman for you?

The answers to these questions became Superhuman's roadmap. They doubled down on what their target customers loved most—speed, keyboard shortcuts, and focused workflow—and built features that would make Superhuman even better: a mobile app, more integrations, and improved attachment handling. In just three quarters, Superhuman's survey results jumped from 33 percent to 58 percent—almost doubling and being far beyond the 40 percent threshold. In 2021, the company raised a financing that valued them at $825 million.[17]

Measuring the strength of your CVP is the strongest single indicator of product-market fit, but it's not sufficient on its own. Let's move on to the next HUNCH metric: Usage Rate.

U—*Usage High: The Toothbrush Test*

Usage refers to how often your target customers use your product or service. The higher the frequency, the better. Google co-founder Larry Page coined a memorable way to think of products that have routine usage: the toothbrush test. When evaluating products to build or acquire, Page would always ask whether the product would be used once or twice a day, like a toothbrush.

The metrics used to measure usage vary widely across industries. Social media platforms and software products often use Daily Active Users (DAUs) and Monthly Active Users (MAUs). Video platforms like Netflix, YouTube, and TikTok not only measure DAUs and MAUs, but also minutes spent viewing content. ClassPass measures usage by the number of classes taken per user.

Usage is a good real-time indicator of product-market fit because you can measure it faster than monthly subscription revenue. As of December 2024, OpenAI reported having 300 million weekly active users (WAUs) on its platform.[18] If that number drops to 290 million next week, internal and external stakeholders would raise the alarm because it could eventually lead to a drop in monthly subscribers. Video platforms like Netflix are keenly aware of their usage rates because a dip could mean more cancellations in the coming months.

High usage is almost always a good thing, particularly when your users are completing what venture capitalist Sarah Tavel calls the "core action" — the key behavior that leads to repeated value for both the user and the business. If your unit economics are off, there may be user value but not business value in that core action. This is what happened to ClassPass when they offered unlimited classes. Power users drove up costs for the company, which hurt their profitability. That is why you should never rely on usage rate as a measure of PMF without also considering the other HUNCH metrics as well.

<u>N</u>—*Net Promoter Score (NPS): Building Word of Mouth*

Many of the most successful startups grow organically with little to no paid marketing. This is often called "going viral" or simply word-of-mouth marketing. Think TikTok, Airbnb, and recently, ChatGPT. OpenAI did not run a massive, coordinated marketing campaign to promote their little chatbot. (I say "little" affectionately; ChatGPT was never meant to be a huge hit.) But once people started using ChatGPT, they couldn't stop talking about it. B2B startups can also go viral; Slack and Notion are two excellent examples.

One way to estimate organic growth is with the net promoter score (NPS). Specifically, NPS measures your customer's willingness to recommend your company to friends and colleagues. The higher your NPS, the more likely you are to grow organically via word of mouth. NPS is measured by asking customers how likely they are to recommend the product on a scale from 0 to 10. After collecting all the responses, you categorize the high scores (9s and 10s) as promoters, the low scores (0–6) as detractors, and the 7s and 8s as passives. You then subtract the percentage of detractors from the percentage of promoters to get your final score. For example, if 20 percent of your customers are detractors and 75 percent are promoters, your NPS is 55 (very good).

NPS can only be measured post-launch, and critics argue that it's not all that useful. After all, what good is measuring a customer's *likelihood* to recommend your product? Measuring actual customer referrals is always preferred, but referral programs are clunky to set up and have innate flaws of their own. That is why I recommend measuring product-market fit across multiple metrics, not just one.

Speaking of flaws, virality isn't a sure sign of startup success either. I bet you can think of multiple "viral" startups that disappeared as quickly as they arrived: Clubhouse, Vine, Fab.com, Groupon, and countless others. Virality should not be the only growth strategy you rely on for success. It's always better as icing on the cake; and ideally, it's built on top of an excellent product and customer experience.

Technically, any net promoter scores above 0 is good because it means you have more promoters than detractors. But startups need as much good word of mouth as possible. I recommend founders shoot for NPS between 40–70—challenging, but doable. For comparison, Amazon is reported to have an NPS of 49 with 66 percent of those surveyed considered promoters and 17 percent considered detractors. That's an excellent score for a $2 trillion company. Netflix's NPS score is 45 as of this writing. The average NPS for companies in the United States is 36 according to Qualtrics.[19]

C—Churn Low: Getting Customers to Stick Around

Churn is an essential metric for a recurring revenue business, such as subscription SaaS companies like HubSpot or membership companies like ClassPass. As a startup searches for product-market fit, churn is an indicator of a product's ongoing ability to meet customer needs. The higher your churn rate, the more customers you are losing each payment period. That means you have to replace those customers just to break even, which puts enormous pressure on your customer acquisition funnel.

The dangerous thing about churn is that it's a *lagging indicator* of customer behavior. Once a customer cancels their subscription, it's too late to win them back. Usage rate is a leading indicator and can often help you predict which way your churn is trending. Churn is best used as a historical analysis tool. I highly recommend measuring churn based on *customer cohort*, or a group of customers that all sign up for your product within the same payment period. Then you can track your churn rate per cohort over time to see the effect of your product changes. If you see your churn rate flattening over time, that's a good sign. (For a deep dive on cohort analysis, visit the appendix.)

Churn rate benchmarks depend on your market. If your customers are large enterprises, your churn rate should be very low—ideally 1 percent or less per month. If your customers are small businesses, only 2 percent or less per month. For consumer businesses, shoot for 3–4 percent or less per month.

In addition, businesses should track both gross churn and net churn. Gross churn measures the percentage of total customers or total revenue lost during a specific period and does not account for revenue gained from existing customers through new or upsell revenue (the numbers I shared above are gross churn). Net churn adjusts for the revenue gained from upsells and new customers. Related to net churn is net dollar retention (NDR), or the total revenue gained from existing accounts.

MongoDB, a database software vendor and Flybridge portfolio company, has an outstanding NDR of 120 percent. That means revenue from existing customers is growing by 20 percent or more. Snowflake has an even better NDR at 130 percent. For both companies, there is a natural growth engine for their customers using more data or more applications—a magical phenomenon also known as "negative churn." This is a sign of extreme product-market fit.

H—*High LTV:CAC Ratio: Perfecting the Business Model*

Let's talk about ClassPass again. After switching to the subscription model with unlimited classes, the startup took off like a rocket. But the business model still had a fatal flaw: Their best customers were costing the company money by taking too many classes. Specifically, these customers were hurting the startup's *unit economics*, or their ability to balance costs and revenue.

You could have the most beloved, high-usage, viral, sticky product on the planet, but if your costs are higher than your revenue, you will go out of business. That is why I tell students and founders that "traditional" product-market fit is necessary *but not sufficient* for startup success. You need an efficient business model, which should be a consideration when evaluating the strength of your PMF.

The final HUNCH metric takes into account the unit economics of your business: Customer Lifetime Value (LTV) and Customer Acquisition Cost (CAC). In short, how much is each customer worth (revenue minus costs), and how much does it cost to acquire them (sales and marketing)? LTV and CAC are typically expressed as a ratio: LTV:CAC.

ClassPass had an LTV:CAC problem. Given their business model construct, power users were driving up the company costs, hurting the LTV (which is a function of profit, not revenue) and forcing the startup to raise prices. This made it harder to attract customers, which increased their CAC. Churn increased at the same time due to higher prices, which hurt LTV even more. This created a negative flywheel effect that would eventually kill ClassPass.

Generally speaking, a healthy LTV:CAC ratio is 3:1 or higher. If a customer is worth $1,500, then you can spend up to $500 on sales and marketing and be comfortably profitable. When ClassPass switched to the unlimited subscription model, their LTV:CAC dropped to 2.8 and was getting worse every day. This metric made it clear to Kadakia that they needed to change the business model.

But there is no perfect metric for measuring PMF—not even LTV:CAC. How you measure lifetime value can dramatically change your ratio and lead you astray. We'll discuss these nuances in Chapter 6.

Benchmarking for Product-Market Fit

The search for product-market fit is a journey, not a one-time milestone. Your focus should be on continuously strengthening your PMF and experimenting with ways to improve each HUNCH metric.

The following are general benchmarks for measuring the strength of product-market fit at each level:

	Nascent	Developing	Strong	Extreme
Hair on Fire CVP (40% test)	20%	30%	40%	50%
Usage High	Varies by company and industry			
Net Promoter Score	40	50	60	70
Churn Low — B2C: / B2B:	8%/Month 5%/Month	6%/Month 3%/Month	4%/Month 1%/Month	2%/Month Neg. Churn
High LTV: CAC	2:1	3:1	4:1	5:1

HUNCH Benchmarks at each stage of Product-Market Fit

Again, these benchmarks are generalized across industries, but they're a good place to start. When you commit to improving each metric through experimentation, amazing things can happen. One of my portfolio companies boasts an NPS score of 84, a twelve-month net dollar retention of 166 percent, and an LTV:CAC ratio of 25x. That is clearly extreme product-market fit! It's no wonder they have received numerous term sheets from top investors in the last year.

Another one of my portfolio company founders recently conducted the 40 Percent Test and proudly wrote to tell me about it. She surveyed her customers and 89 percent of respondents indicated they would be Very Disappointed if they couldn't use her product anymore—more than doubling the 40 percent threshold set by Sean Ellis.

The search for product-market fit has not changed in the age of AI—it has only accelerated. 10x Founders have the tools to build, measure, learn, and iterate faster than ever. This fact makes your experiment selection all the more crucial.

Whether you're building B2B SaaS or rocket ships, your journey to PMF and startup success starts with nailing your customer value proposition.

Chapter 4

Hair on Fire: Discovering Your Customer Value Proposition

LET ME REITERATE MY definition of a customer value proposition from Chapter 3:

> *A customer value proposition (CVP) is your unique solution to a painful problem for a specific target customer. It's WHO you serve, WHAT problem you solve, and HOW you solve it.*

One way to measure the strength of your CVP is with the 40 Percent Test: If you asked your customers, "How disappointed would you be if our product disappeared?" What percentage would be "Very Disappointed?" Anything over 40 percent indicates you've discovered a hair-on-fire problem and effective solution.

To discover and refine your CVP, founders need to continuously test three hypotheses:

- ***WHO hypothesis***: Who is the ideal customer? Founders often struggle to narrow their initial focus to a single, well-defined, well-articulated customer segment. My advice: pick a narrow customer definition, and then go even narrower.

- ***WHAT hypothesis***: What is the target customer's underserved need— their "hair-on-fire" problem? The answer will help you refine the product features and capabilities of your MVP.

- ***HOW hypothesis***: How will you solve the customer's problem with technology, service, and operations? (And a focus for later in the process: How will you deliver this solution profitably?)

The order in which you tackle these hypotheses will differ from startup to startup. Remember our criteria for test selection: Which hypothesis is most controversial? Which will get you to the next fundraising milestone? Which represents the greatest risk?

Squire and ClassPass both started with WHAT hypotheses. The founders faced *hair-on-fire* problems in their everyday lives: For Payal Kadakia, it was the pain of finding a fitness or dance class. For Songe LaRon and Dave Salvant, it was the frustration of getting a haircut. They set out to create products that made *their* lives easier, what many people would call "scratching your own itch." Their WHAT hypotheses were validated early on with consumer-facing minimal viable products (MVPs). They knew they had found problems worth solving.

Squire and ClassPass then moved on to their WHO hypotheses. Squire's original WHO hypothesis was wrong. They thought barbershop customers (like them) were the target customer and that they'd charge them a 1 percent transaction fee. It turned out that the real customers were barbershop *owners* who struggled to manage their businesses with existing tools. This target market represented a much larger opportunity for the startup. ClassPass's WHO hypothesis was only half right. The founders were right about targeting fitness class enthusiasts, but they also had to expand their WHO to include gym owners in what became a two-sided marketplace business model.

However, both startups struggled for years to nail the HOW hypothesis, testing several iterations of their products and business models before discovering in-demand, profitable solutions.

This is the most common path for startups in search of a strong CVP: WHAT, WHO, HOW. But it's not the only path. C16 Biosciences took a different approach. They also started with their WHAT: a massive, *world-on-fire* problem. But then they turned their focus to the HOW, using science to develop an innovative solution. Now, they are in search for the right customer—their WHO.

C16 Bio: Developing a CVP for a Science-Based Startup

C16 Biosciences emerged from a class project at MIT in 2016.[20] Co-founders Shara Ticku, David Heller, and Harry McNamara connected over a shared passion for using synthetic biology to tackle climate challenges. As Heller recalls, "The class inverted the typical technology application process—rather than create a technology and look for potential applications, we were instructed to look for a problem, then find the technology to solve it."

The problem that ultimately captured their attention was the environmental impact of palm oil, a highly versatile vegetable oil extracted from palm trees, which flourish in the tropics. Palm oil is used as the primary cooking oil in many developing nations, as well as an ingredient in various food products, detergents, cosmetics, and even biofuel.

Palm oil is relatively cheap to produce, which is a reason why it's so popular. Unfortunately, palm oil plantations have expanded rapidly at the expense of tropical forests, harming local communities that depend on them and wiping out vital habitats for endangered species. Ticku witnessed this firsthand during a trip to Singapore in 2013, where she had to wear a mask while outdoors because of the haze from land-clearing operations in nearby Indonesia for palm oil plantations. The trio had found a problem that "fell directly in the sweet spot of providing a challenging biological problem and solving for a well-established need in the market," they wrote in their Y Combinator accelerator application. Synthetic alternatives to biological products, such as lab-grown meat, were becoming popular, and investors were eager to pursue this new frontier. Armed with some technical knowledge (Heller was an undergraduate at MIT studying bioengineering and McNamara was a physics PhD student with experience in synthetic biology) and passion for the problem, the founders decided to engineer a microbial alternative to palm oil. "It's such a big industry . . . if we can replace just 5 percent of the market—or 3 million metric tons—we could stop the release of 1.5 gigatons of carbon each year into the atmosphere," said Ticku.

The founders secured early funding from strong showings at business plan competitions like Harvard President's Innovation Challenge and MIT Sandbox Innovation Fund, each of which awarded them $25,000, enough to conduct proof-of-concept experiments. "These business competitions validated the technology and the problem, convinced us to keep working on the company, and provided a little bit of money so we could actually do experiments," explained Ticku.

The problem was validated, but the science was not. While biologically possible, the method of creating synthetic palm oil was not a known solution. The team pursued a cost-effective approach involving fermentation, similar to brewing beer. Still, the process took nearly two years to get right.

Beyond the biological challenges of creating a synthetic palm oil, the founders faced another problem: narrowing down their target customer. Because palm oil was so ubiquitous, they could credibly claim that "everybody is our customer." One day that may be true, but C16 Bio needed a *beachhead* customer—a narrow slice of their total addressable market—to focus their efforts. The perfect beachhead customer would have the right mix of hair-on-fire need for their synthetic oil and the stomach to take a chance on an unproven startup. They would also need to be relatively price insensitive since C16 Bio's product would be more expensive than traditional palm oil at first. The team considered various potential customers, including growers, chemical companies, food brands, personal care brands, and end consumers in the personal care market.

Through extensive research and customer interviews, C16 Bio narrowed down its choices to personal care brands and the food industry. Personal care brands were less price sensitive due to higher industry margins. They were also more innovative and willing to try new things; being associated with sustainable practices mattered to these brands. Finally, the relatively small amount of palm oil used in personal care products meant the startup could gradually ramp up their production.

The food industry was appealing for different reasons. First, it was massive, and demand would be higher long-term. Plus, food brands were already reaching out to C16 Bio and showed significant interest at trade shows. Food manufacturers would also need larger quantities of palm oil, meaning much larger accounts for the startup.

In addition, C16 Bio was still getting interest from potential customers in other industries. This wave of demand created another challenge for the startup: Every potential customer wanted samples, but they couldn't create the product fast enough. Unlike software companies that could rapidly prototype a minimal viable product, "there are no hacks in fermentation," Ticku explained. "Biology has a different pace than bits and bytes." Heller echoed the frustration: "The demand was clearly there from so many different angles, but the science wasn't moving fast enough."

Signals from both food and personal care brands were promising, but C16 Bio decided to target personal care as their beachhead market. The fact that personal care brands needed *less* palm oil was appealing to the founders. They could refine their product and operations without sinking capital costs into scaling up to meet demand from food companies and other higher-volume customers.

In 2021, C16 Bio secured a $20 million Series A round led by Bill Gates' Breakthrough Energy Ventures and continued to grow and refine their operations. In 2023, they launched a direct-to-consumer cosmetics brand called Palmless and partnered with personal care brands to produce palm-free skincare products. More recently, they've begun to partner with food brands to test their product as an alternative to the oils used in non-dairy ice cream, cheese, and baked goods.

C16 tackled the CVP hypotheses in order of risk and opportunity: First they validated the problem (the WHAT), developed the solution (the HOW), and finally identified the right initial customer (the WHO).

Your approach may be different. Most startup journeys look more like ClassPass and Squire: Identifying a painful problem (WHAT), choosing a target customer (WHO), and then developing a solution (HOW). Let's look at each step of the CVP experimentation in that order, and how AI can accelerate the process.

WHAT Hypothesis: Uncovering "Hair-on-Fire" Problems with Research Driven Ideation

The genesis of any successful startup is a deep understanding of market needs, pain points, and the competitive landscape. My colleague, Scott Brady of Stanford, thinks of this process as a research exercise and has coined the term Research Driven Ideation (RDI).

RDI helps entrepreneurs gain a 360-degree view of an industry through in-depth interviews with experts and rigorous data analysis. The process feels a bit like a graduate project, systematically researching promising market segments, spotting gaps, identifying non-obvious problems, and uncovering opportunities for disruptive solutions.

C16 Bio is an example of a startup idea developed through RDI. The founders were not avid palm oil users or industry experts. They sought out a problem to solve and did their homework to develop a viable solution.

The RDI process unfolds in three phases:

1. Team formation and alignment: Select research partners, define success criteria, and create an opportunity rubric which outlines the ideal industry attributes to explore.

2. Bench-level analysis: Choose an initial focus area, become "cocktail party conversational" in the space, develop a market hypothesis, and identify interview targets.

3. Research, synthesis, and evaluation: Conduct curiosity-driven conversations with industry insiders, iteratively gather more knowledge, test hypotheses, and evaluate potential opportunities.

RDI doesn't guarantee success, but it dramatically increases the odds of zeroing in on an idea that truly matters.

Generative AI can be a powerful tool to assist in RDI. In many ways, it is the ultimate research partner:

Comprehensive Market Analysis: Gen AI is brilliant at summarizing, synthesizing, and recognizing patterns. For this reason, the first assignment in my Launching Technology Ventures (LTV) class at HBS is for students to use ChatGPT to come up with an original startup idea. They use it to analyze vast amounts of data from market research reports, online forums, and social media to identify unmet needs and emerging trends.

Competitive Intelligence: Because ChatGPT and other LLMs (like Google's Gemini) are always crawling the web and ingesting publicly available data, they are masters of competitive intelligence. They can gather and synthesize information on competitors or product substitutes, then identify gaps in their offerings and potential areas for differentiation. There are now AI tools specifically for competitive monitoring and analysis—effectively Google Alerts on steroids—tracking personnel changes, marketing launches, traffic, and other publicly available data.

Problem Discovery: Finally, AI is an excellent brainstorming partner for problem discovery. As an illustration, let me walk you through the class assignment I give my LTV students on day one. First, they pick a specific sector and general market, for example, B2B payments in Brazil or health IT for Medicare Advantage patients.

Then they ask ChatGPT to act as a co-founder and thought partner to brainstorm hair-on-fire problems. Here is a sample prompt they might feed into ChatGPT:

Hair on Fire: Discovering Your Customer Value Proposition

> I am an aspiring founder and I want to come up with a new, compelling startup idea that will lead to the creation of a valuable, enduring company. The area I'm most interested in is the Medicare Advantage market. I believe that software, automation, and AI can be applied to make this market far more efficient, benefiting everyone in the system. Act as my co-founder to help me ideate and come up with five new startup ideas in this sector. For each one, tell me the target customer and describe a "hair-on-fire," "must-have" customer value proposition.
>
>

Here's how ChatGPT responded in this instance (your responses will vary as these models are non-deterministic and ever-evolving):

ChaptGPT:

Great! Let's brainstorm some innovative startup ideas focusing on the Medicare Advantage market, leveraging software, automation, and AI:

1. **AI-Driven Personalization Platform**
   - **Target Customer:** Medicare Advantage plans.
   - **Value Proposition:** This platform uses AI to personalize healthcare plans and services for individual beneficiaries, significantly enhancing user satisfaction and adherence to care plans, thus reducing overall costs.

2. **Automated Claims Processing System**
   - **Target Customer:** Medicare Advantage plans and healthcare providers.
   - **Value Proposition:** Automates and streamlines the claims process, drastically reducing processing times and errors, improving cash flows, and increasing operational efficiencies for providers.

3. **Predictive Analytics for Patient Risk Scoring**
   - **Target Customer:** Medicare Advantage plans.
   - **Value Proposition:** Utilizes advanced predictive models to identify high-risk patients early, enabling proactive care that can prevent costly interventions and hospitalizations, thereby saving on costs and improving patient outcomes.

4. **Virtual Health Assistant**
   - **Target Customer:** Medicare beneficiaries.
   - **Value Proposition:** A 24/7 AI-powered virtual assistant that provides immediate health advice, medication

reminders, and scheduling assistance, greatly enhancing accessibility and adherence to health regimens.

5. **Fraud Detection AI**
   - **Target Customer:** Medicare Advantage plans.
   - **Value Proposition:** An AI tool that detects and prevents fraudulent claims in real-time, saving millions in potential losses, and increasing trust and efficiency in Medicare claims processing.

Let's say you like one of these ideas—fraud detection—and want to explore it further. Prompt ChatGPT to help you dive deeper into this idea. Perhaps you want assistance with customer discovery and to design an experiment to test this value proposition. Here's a prompt to help:

The fifth idea, fraud detection, is really interesting. Describe that idea in more detail and suggest two actions I might take to do deeper customer discovery and an experiment I might run to test the value proposition to ensure that it is truly a "must-have" and not just a "nice-to-have."

ChatGPT:

Concept: Develop an AI-driven software solution that integrates with existing Medicare Advantage plan systems to identify and prevent fraudulent activities in claims processing. The software would use machine learning algorithms to analyze patterns and anomalies in claims data, flagging potential fraud for further investigation.

How It Works:
- **Data Collection:** The software aggregates claims data from various sources within the Medicare Advantage ecosystem.
- **Pattern Recognition:** Machine learning models are trained on historical claims data, learning to identify patterns typical of legitimate claims and those suggestive of fraud.
- **Real-Time Analysis:** As new claims come in, the system analyzes them in real-time, comparing them against learned patterns.
- **Alert System:** Suspicious activities are flagged and categorized by potential risk level, notifying compliance officers or auditors within the organization.
- **Feedback Loop:** The system continuously learns from the outcomes of flagged claims, improving its accuracy and reducing false positives over time.

Customer Discovery Actions

1. **Stakeholder Interviews:**
   - **Action:** Conduct detailed interviews with compliance officers, claims processors, and financial auditors within Medicare Advantage plans. These interviews will aim to understand their current challenges with fraud detection, the impact of fraud on their operations, and their requirements for a solution.
   - **Goal:** To gather insights into the specific pain points and operational needs that our solution must address to be considered valuable.

2. **Industry Analysis:**
   - **Action:** Analyze industry reports and conduct secondary research to understand trends in Medicare fraud and current solutions in the market. This may include reviewing case studies of fraud and how other healthcare programs are managing risks.
   - **Goal:** To position our solution appropriately in the market, ensuring it is not only competitive but also superior in addressing unmet needs.

Experiment to Test the Value Proposition

Pilot Program:

- **Setup:** Collaborate with a willing Medicare Advantage plan to implement the AI fraud detection system on a segment of their claims data. The system would run in parallel with their existing processes, without disrupting them.
- **Metrics:** Monitor several metrics such as the number of fraud cases detected, the accuracy of the detections (measured by false positive and false negative rates), the time saved in detecting fraud, and the reaction of the users to the tool's interface and alerts.
- **Objective:** To demonstrate the system's effectiveness in detecting fraud more accurately and efficiently than current methods, and to show that it can significantly reduce the costs associated with fraud investigations and losses.
- **Feedback Collection:** Regularly gather feedback from the users on the usability of the system and its impact on their daily operations.

This pilot would serve as a crucial litmus test for the value proposition, proving that the system is not only innovative but provides tangible, must-have benefits to Medicare Advantage plans.

When I had my 250 students run this experiment in 2024, they were shocked by how effective ChatGPT is as a co-founder and thought partner. Their experience has been validated by a study done by Wharton and Cornell professors. In their paper titled, "Ideas Are a Dime a Dozen: LLMs for Idea Generation in Innovation,"[21] they showed that ChatGPT 4 was able to generate higher-quality startup ideas on average as compared to Wharton students. They also showed that the vast majority of the best ideas in the pooled sample came from ChatGPT 4.

After this relatively simple assignment, I ask my students to go even further in their relationship with AI. I ask them to create a customized GPT called "GPT Co-Founder" and train that co-founder on their target sector. Creating a customized GPT in ChatGPT takes less than an hour. Here is the protocol:

- Explore a series of startup ideas in a particular domain/sector of interest (using the prompts I just shared).

- Create a custom GPT in ChatGPT and prompt it to be your AI co-founder who is a domain expert and knows the target user persona intimately (suggest it has had a decades-long career in the field).

- Create a second custom GPT that represents your target customer persona (more on this below).

- Conduct customer discovery with the help of your AI co-founder and AI customer persona.

- Define your WHAT hypothesis and ask your AI co-founder for feedback.

- Ask your AI co-founder to design an experiment to test the hypothesis.

Think about the implications of such a workflow: months of problem discovery and research driven ideation completed in a matter of hours. No, AI cannot replace interviews with real customers, but you can trust it to jumpstart your knowledge on a new market and provide insight on both good and bad ideas. Imagine if LaRon and Salvant had ChatGPT to research the barbershop industry. How much time would they have saved? We'll never know the answer to that specific question, but founders today would be foolish not to use AI as a co-founder during the ideation phase.

WHO Hypothesis: Identifying your Ideal Customer Persona

Many founders start with a problem they've experienced in their own life. They represent the early ideal customer profile, or ICP, but this persona often needs to be revised or expanded, as was the case with Squire and ClassPass. In some cases of science or technology-driven startups, like C16 Bio, identifying the ICP is a more formal search process. Either way, customer discovery is a critical, and often overlooked, piece of developing a strong CVP. Again, gen AI can help you get the ball rolling:

Analyze Customer Data: Founders can use AI to process customer reviews, feedback, and behavior patterns, building a nuanced understanding of customer preferences and expectations.

Create Hyper-Specific Customer Personas: You can then use AI to refine and bring to life customer personas, tailoring product features and marketing strategies to match identified customer segments. AI-based personas give you and your team on-demand "customers" to ask questions and get feedback. Some key questions to explore in the customer discovery process:

- What does a day in this customer's life look like?
- What tasks do they perform and what tools do they use?
- Where do they experience the most friction and frustration?
- What would make their life or work easier, more productive, or more enjoyable?
- How are they evaluated and compensated in their work role?
- Who do they report to and who are their key stakeholders?

One of my colleagues, HBS Professor Ayelet Israeli, conducted research on the use of ChatGPT to respond to customer surveys and found its answers were strikingly realistic and consistent with well-documented patterns of consumer behavior. In other words, you can train AI to mimic actual customer research, even accurately estimating willingness to pay.

To continue the example about an AI-native fraud detection platform for Medicare Advantage, an aspiring founder might ask ChatGPT to play the role of a claims processor (the potential ICP) and answer some of the questions posed above:

> Pretend you are in charge of claims processing at a Medicare Advantage plan. I want to interview you as if you were a target customer for my fraud detection AI startup so that I can do a deeper customer discovery and needs assessment. Every time I ask you a question for the next three questions, answer as if you are this user persona.
>
> First question: What does a workday look like for you? Walk me through the main tasks you perform and what tools you use to perform them.
>
>

ChaptGPT:

As the manager in charge of claims processing at a Medicare Advantage plan, my day typically involves overseeing the workflow of claims submissions, processing, and approvals to ensure everything runs smoothly and efficiently. Here's a breakdown of my main tasks and the tools I use:

1. **Reviewing Claims Submissions:** Each morning starts with reviewing the new claims that have been submitted by healthcare providers. I check for any that are flagged by our preliminary screening tools for errors or inconsistencies.

2. **Claims Adjudication:** I spend a significant part of my day on claims adjudication, which involves assessing claims for their validity and compliance with our plan's coverage rules. For this, I use our claims management software, which integrates data from various sources and helps identify discrepancies.

3. **Team Coordination:** I frequently meet with my team to address complex cases, discuss any issues they're encountering, and provide guidance on processing specific types of claims. It's crucial to maintain high accuracy and compliance with regulations.

4. **Using Analytics Tools:** I use analytics tools to generate reports on claims processed, turnaround times, and error rates. These reports help us identify areas for improvement in our processing techniques and training needs for the team.

5. **Liaising with Other Departments:** I coordinate with the compliance and fraud detection teams to discuss any suspicious claims and follow up on investigations. We use a basic fraud detection system that flags anomalies, but it often requires manual intervention to verify.

6. **Provider Interaction:** Part of my role involves communicating with healthcare providers to clarify claim details or inform them of denials or adjustments. This ensures that there are clear lines of communication and helps reduce disputes. The tools we rely on include claims processing software, electronic health records (EHR) integration, basic analytical tools, and a preliminary fraud detection system. While these tools are integral to our operations, they do have limitations, especially in efficiently detecting sophisticated fraud schemes without a lot of manual input.

This example shows what you might do if you were to use the public ChatGPT tools. There are also specialized tools for user persona research. For example, Sprig creates AI personas that can serve as "customer doubles," responding to new product or feature ideas and highlighting pain points to inform the product roadmap. They even allow companies to capture user behavior video clips and automatically turn the observed behavior into product insights. Another company, Insight7, ingests actual customer interview recordings and conversations from a range of sources (including HubSpot and Zendesk) and extracts insights and sentiment that you can use to sharpen marketing and sales efforts. There are also dozens of startups that will build synthetic focus groups or AI customer panels for you.

Tools like these are just getting started in their ability to assist founders in their customer discovery and ideation process. Armed with this AI-powered insight, real customer interviews can be more nuanced and informed.

HOW Hypothesis: Building Your MVP

No startup idea survives first contact with actual customers. All the research in the world is useless until you get a working prototype into the hands of your users.

Your minimal viable product (MVP) is a basic version of a product with only the most essential features that your ICP *needs* to have. No bells and whistles—just the features needed to evaluate your customer value proposition. Your goal is to answer a few key questions:

1. Is your product essential? In other words, does it address a hair-on-fire problem?
2. Is your ICP willing to use and pay for your product?
3. Can you solve the problem in a profitable way?

AI is an incredible tool for rapidly developing and testing MVPs—and it's so much easier to do than you might expect. I recently taught a hands-on workshop to my Launching Tech Ventures class on using AI tools to catalyze prototyping. Two-thirds of the students had no technical background, yet in a few hours, all of them had working MVPs of an idea "coded" entirely from scratch.

The workshop was designed to force the students to use the modern AI tools to transform a startup idea into a Lean Canvas business model (a popular, one-page business model template), and then a prototype. The exercise had two parts:

First, students used an AI business model generator called SparkRockets to pressure test their startup concepts. This tool, which we selected after evaluating several options, helps founders quickly generate and iterate on their Lean Canvas.

Next, students built a landing page from scratch using ChatGPT, Visual Studio Code, and GitHub Pages. While tools like SparkRockets can auto-generate landing pages, we wanted students to start with a blank screen and "program" in HTML using English as their programming language. This assignment gave them a baseline understanding before using AI tools to refine their pages.

Amazingly, the biggest barrier for students was their imagination. "Tech is no longer the moat!" said one student who finally broke through the mental block. The hardest part was discarding their preconceived notions of what was possible. Once they began using these tools, they saw how simple and intuitive they were.

SparkRockets is one of many tools that provides some shortcuts for building your MVP, and more are coming. But in truth, you don't always need a specialized AI tool. ChatGPT—and the other general purpose chatbots—continue to improve and can be very effective at helping you define what your MVP should look like.

Let's return to the AI-native fraud detection platform for Medicare Advantage as an example:

> Based on the interview insights about the claims manager's workflow and pain points, suggest the key features for an MVP fraud detection platform focused on reducing manual review time and more effectively catching sophisticated fraud.
>
>

ChaptGPT:

Based on the insights from the claims manager interview, here are the key features I suggest for an MVP fraud detection platform:

1. **Advanced anomaly detection:** Utilize machine learning algorithms to identify patterns and flag suspicious claims more accurately, reducing false positives and manual review.

2. **Real-time alerts:** Implement a real-time notification system to alert the team about potential fraud cases as they occur, enabling faster response times.

3. **Seamless integration:** Ensure the platform integrates smoothly with existing claims management systems and EHRs, minimizing disruption to workflows.

4. **Intuitive dashboard:** Provide an easy-to-use dashboard with visualizations of key metrics, flagged cases, and team performance to help managers make data-driven decisions.

5. **Continuous learning:** Incorporate a feedback loop where the system learns from each reviewed case, improving its accuracy over time.

Then, review this list of features and develop the detailed product requirements behind them. ChatGPT can once again help (ChatPRD, the product development tool I mentioned in Chapter 2, is tailor-made for this task). Finally, cross-check with your own experience and research. Your MVP should only include *need-to-have* features for your narrowly defined target customer. All other features should be added to your roadmap and left untouched until you've validated your core concept.

Once you have built your MVP—whether it's a landing page or basic working app—it's time to test it. Your AI co-founder can help you design this experiment:

> Suggest an MVP experiment to test if the AI fraud detection platform actually saves claims managers significant time and reduces fraud compared to their current process. Provide specific metrics to measure.

ChaptGPT:

To test the effectiveness of the AI fraud detection MVP, I suggest running a controlled experiment:

1. Identify a pilot group of claims managers to use the MVP platform and a control group to continue using the existing system.

2. Over a set period (e.g., three months), track the following metrics for each group:
   - Average time spent on manual claim reviews
   - Number of fraudulent claims detected
   - Percentage of claims flagged as potentially fraudulent
   - False positive rate (claims flagged as fraudulent that were legitimate)

3. Conduct user satisfaction surveys with the pilot group to gather qualitative feedback on ease of use, perceived time savings, and effectiveness.

4. Analyze the data to compare metrics between the pilot and control groups. The MVP platform would be considered successful if it demonstrates:
   - Statistically significant reduction in average manual review time
   - Increased detection of fraudulent claims
   - Reduced percentage of claims requiring manual review
   - Lower false positive rate
   - Positive user feedback and satisfaction scores

If the MVP achieves these results, it validates the value proposition of saving time and improving fraud detection accuracy. This data can inform further iterations and refinements before a full-scale rollout.

What's fascinating to me about this answer is that ChatGPT gives you key performance indicators (KPIs) to validate your experiment. Based on my experience, this is a good list of metrics to track. I would also add the 40 Percent Test to track.

Founders today can dramatically accelerate their search for a strong CVP and validate their WHO, WHAT, and HOW hypotheses. But it's still important to do things that *don't* scale, even in the age of AI.

Find 100 People Who Love You

Paul Graham of Y Combinator made the phrase, "Do things that don't scale" famous, and for good reason. Brian Chesky of Airbnb called it, "The best piece of advice I ever got." But what does it mean, exactly?

Chesky provides a helpful reframe: "It's better to have one hundred people *love* you than a million people sort of like you."

In other words, founders need to start by completely satisfying a small number of customers before trying to scale. This is why it's so important to nail your customer value proposition. You need a real "hair-on-fire" problem to solve, and then a product that becomes a necessity in your customers' lives. If you have yet to reach the 40 Percent Test threshold, keep iterating on your WHO, WHAT and HOW. You are not ready to scale.

The Squire founders diligently focused on their CVP in the early days, going to extreme lengths to serve their customers. Purchasing a barbershop is a quintessential example of doing things that don't scale; it gave LaRon and Salvant a laboratory to experiment and build their barbershop operations platform.

One of Intuit's founders, Scott Cook, famously asked customers if he could follow them home to watch them use his flagship product, Quicken. A little forward? Yes. Scalable? Absolutely not. But it gave Cook irreplaceable feedback that still benefits the company today. In fact, Intuit institutionalized the practice, making it a regular part of their Product development process. The Follow Me Home program proves you're never too big to do things that don't scale.

AI or no AI, your best insights will still come from doing things that don't scale. You can accelerate your learning curve and experimentation rate, but don't use AI as a substitute for getting your hands dirty.

Chapter 5

Laws of Attention: Go-to-Market in the Age of AI

First-time founders are obsessed with product. Second-time founders are obsessed with distribution. —Twitch co-founder Justin Kan

GO-TO-MARKET (GTM) USED TO be pretty simple: Spend, baby, spend! Buy ads, hire a team of sales reps, and spend thousands of dollars on a PR stunt. Hell, just *give away money* if it means you'll sign up more users than the startup next door (as was literally the case of PayPal vs. X.com in the early 2000s). Of course, I'm being a bit facetious, but this was prevailing wisdom for decades; the startup with the largest war chest would win on sheer distribution power.

Today, all that has changed. Interest rates are no longer zero and the cost of capital is much higher. Investors are more concerned about customer acquisition costs and unit economics. Startups need to stay lean, and founders need to be creative in how they reach, activate, and convert customers. It just so happens that generative AI is tailor-made for this sort of thing.

Your startup's GTM strategy is a mix of positioning and messaging, market selection, demand generation, distribution channels, sales funnels, and strategic partnerships—all working together to turn cold prospects into loyal, evangelical customers as cost-effectively as possible. These activities fall right into the sweet spot of generative AI as we know it today: content creation, lead nurturing and engagement, personalization, and data analysis. Incredibly, just as startups around the world were being forced to tighten their belts, they were given this magical new technology that stretches their GTM dollars further than ever.

In this chapter, we will walk through the principles of go-to-market strategy and how to use AI to 10x your growth potential in an efficient manner. We will also cover the four key hypotheses every GTM strategy must test:

- Initial Market hypothesis
- Growth hypothesis
- Sales Model hypothesis
- Partner and Channel Choice hypothesis

Let's start by discussing a relevant case study: a startup that began life immediately before the gen AI revolution and is now shifting tactics to take full advantage.

AllSpice: Building the GitHub for Hardware Teams

In the U.S., technology hardware companies lose $35 billion a year on mistakes and rework—the equivalent of 125 million hours of engineering time—due to poor version control.

In software development, version control means tracking edits and changes to a codebase from multiple developers. Virtually every software team on the planet uses a version control tool like GitHub, GitLab, or Bitbucket to coordinate work and protect their production systems.

Hardware engineers need version control systems as well. Every piece of hardware starts as a digital plan, often a 2D model built in a computer-aided design (CAD) system. But unlike the software world, there is no version control tool for the hardware industry. Large companies implement layers of red tape and employ entire teams of people to manage versioning. Small teams try to use tools like GitHub, but it's like trying to put out a fire with a brick. "GitHub was ultimately designed for text-based software files, not the 2D circuit data we worked with," said AllSpice co-founder Kyle Dumont, an electrical engineer by training. "GitHub would not take a CAD design and visualize it."[22] This meant teams couldn't easily see the differences between versions of the same design, leading to numerous costly mistakes and rework—$35 billion dollars' worth every year.

AllSpice aims to fix this by building the GitHub for hardware engineers. Dumont and classmate Valentina Ratner co-founded AllSpice in 2018 while at HBS. Ratner, a software engineer, was interested in building tools for hardware teams. Dumont had experienced all the pains of hardware version control while working at iRobot.

The opportunity to solve the version control problem for hardware companies is massive. In the software world, GitHub was purchased by Microsoft in 2018 for $7.5 billion, while GitLab went public and reached a $15 billion market cap on its first day of trading. With over 300,000 electrical engineers working in 30,000 hardware companies, Ratner and Dumont estimated the total market opportunity to be over $5 billion to start.

After graduating from HBS in 2020, Ratner and Dumont went full-time on AllSpice—Ratner serving as CEO and Dumont the CTO. They entered the HBS New Venture Competition (NVC) and made it to the finals, which earned them important publicity early on. They released a beta version of AllSpice that summer and landed a few hundred early customers. These early customers came primarily from the founders' networks, as well as press coverage from NVC.

Inbound demand was never an issue with AllSpice. As a team collaboration tool, AllSpice was inherently viral: users who liked the product would naturally invite their teammates to join them. Ratner and Dumont worked closely with their beta users as design partners to create the features and tools they needed most. They saw early signs of product-market fit: top cohorts of users were spending nearly forty hours per month on the platform with over one hundred interactions per week.

After a successful beta period, AllSpice launched their first public product in January 2022. Inbound interest flooded in and the team saw ravenous demand for the product. "I was doing conversations with community users, then getting on calls to do enterprise deals with IT and procurement, and then working with self-service inbound requests. I was doing a lot of things, and none of them very well, so I had to focus."

AllSpice had to decide on a GTM strategy: Do they prioritize self-service freemium users, small-to-midsize teams, or enterprise clients? With their natural inbound demand, freemium seemed like the obvious route, and AllSpice's investors backed that plan too. The popular GTM playbook at the time was to amass as many users as possible and figure out monetization later (a strategy we'll discuss again in Chapter 6). The board of directors pushed Ratner to spend more aggressively to grow fast.

But just months after AllSpice's public launch, the global markets shifted, and so did the opinions of AllSpice's board. The U.S. Federal Reserve raised interest rates for the first time in three years. Suddenly, the $3.8 million Seed round AllSpice had raised didn't seem like nearly enough. The board cautioned prudence, and Ratner had to come up with a new growth plan—one that preserved cash and promised better unit economics.

AllSpice made the difficult choice to end the free plans it started during its beta period. While their customer acquisition cost was near zero, the lifetime value of free users was too low to justify their focus. However, the team kept aspects of the freemium model by allowing "collaborators" to join projects for free (AllSpice would charge for "committers," or users who committed final versions of the code to production). This pricing plan incentivized collaboration and word-of-mouth growth within teams.

The founders eventually chose to target small-to-midsize teams, especially startups. "We started with startups because they are very innovative and move quickly, which are values that we encompass," said Ratner. With inbound demand still strong, Ratner built a small team of junior sales reps to run demos and close deals. They were growing steadily, and most importantly, learning from their early customers along the way.

But Ratner never took her eyes off the big prize: enterprise sales. Given the market environment and thin balance sheet at the time, AllSpice couldn't afford to build an enterprise sales team with its expensive account executives and long sales cycles. Instead, Ratner's team took a "land and expand" approach by attracting individual engineers and small teams from within large organizations and then growing organically. They didn't need to flip an entire enterprise at once; they could grow *bottom-up*.

AllSpice's small GTM team was making the most of their limited resources, yet Ratner still felt capped. She knew they had strong demand and growing product-market fit, but they didn't have the runway to scale more aggressively.

Then everything changed again in late 2022, with the launch of ChatGPT.

AI for GTM: Personalization at Scale

Before November 2022, AllSpice was forced to make a decision that all companies in the past have had to make: Should they target large customers with long, expensive sales cycles, or smaller customers that are easier to acquire but less valuable? This tradeoff is illustrated by the chart created by venture capitalist Christopher Janz:

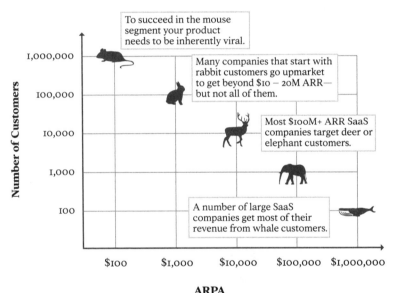

Customers that represent a lower average revenue per account (ARPA) are generally easier to acquire and serve, but you need many more of them to build a sizeable company. The largest accounts—the elephants and whales—require a lot of handholding and personalized attention, making it costly to sell to and support them.

AI is beginning to break this curve. Suddenly, small companies like AllSpice are able to sell to enterprise-level clients with personalized outreach, content, and communication at scale.

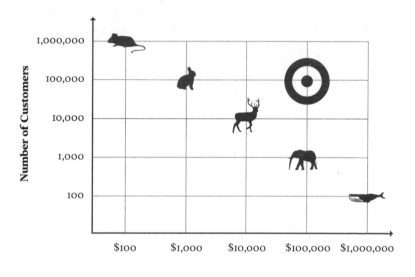

ARPA
(Average Revenue Per Account per year)

AllSpice is brilliantly using AI to nurture inbound leads—a time- and resource-intensive process. The startup's small sales team is using tools like ElevenLabs and HeyGen to create thousands of customized demonstration videos based on a single recording. Not only is this radically more efficient, but AllSpice's customers—engineers themselves—often prefer it over having a Zoom call with a salesperson. They also use a sales execution tool called Outreach to automatically sequence outbound emails and follow up on tasks to move prospects through their pipeline. Outreach recently incorporated gen AI to help sales reps understand true buyer sentiment, summarize deals, generate personalized emails, respond to prospects, and forecast pipeline more accurately. These new workflows have freed up Ratner and her sales team to focus on the largest, most valuable clients while consistently converting small and midsize teams.

AI has helped to scale AllSpice's customer support experience as well—a critical aspect of servicing large companies and enterprises. Ratner's team is using generative AI to automatically generate video tutorials and documentation for every feature in their toolset. Crucially, AI helps them keep this documentation up to date as the AllSpice product evolves rapidly.

In addition to sales and support, we are seeing an explosion of AI tools across the go-to-market spectrum from sales development reps (11x) to sales engineers (DocketAI) to customer data enrichment (Clay) to personalized videos (HeyGen) and more. We'll discuss more specific AI use cases in the GTM hypotheses sections later on.

To be clear, AI will not make the difficult strategic decisions for you: who to target, how to sell them, and your best channels for growth. But it can be a fantastic aid in your search for these answers and a supernatural sidekick to scale them.

The Go-to-Market Hypotheses: Four Questions to Answer

In the early stages of a startup's life, the primary objective of GTM experiments is to discover a repeatable and scalable process for customer acquisition. There are four key questions to answer:

1. What is my initial market?
2. What is my growth and demand generation strategy?
3. What is my sales model?
4. Who are the best (if any) channel partners?

The result of successful GTM experiments is a playbook that leads your team to consistent customer acquisition while maintaining favorable unit economics—i.e., lifetime value to customer acquisition cost (LTV:CAC) ratio, ideally at or above 3:1 (as noted in the HUNCH framework in Chapter 3).

But even when you answer these four questions, there is another decision to make: How quickly should you scale? When do you "pour fuel on the fire" to grow quickly? The answer depends on the strength of your product-market fit and the market conditions. You need to be aware of your next fundraising milestone and the momentum needed to get there. On the other hand, one of the greatest risks for a startup is scaling prematurely. Many startups have tried to force growth despite insufficient PMF and poor timing. Especially in an environment where capital is scarce, keeping "dry powder" and extending your runway is generally wise. Luckily, thanks to AI, you can do so much more with less. We will discuss scaling your GTM strategy in detail in Chapter 8.

Now let's talk about the four GTM hypotheses.

Initial Market Hypothesis: Targeting Innovators

In Chapter 4, we covered the WHO hypothesis: identifying your ideal customer profile (ICP), or *beachhead* customer. Tied to this concept is the Initial Market hypothesis. Think of the relationship this way: Your ICP "lives" in your initial market. Squire's ICP was barbershop owners, and their initial market was independent barbershops (as opposed to hair salons or barbershop chains). AllSpice's ICP was electrical engineers, and their initial market was small-to-midsize engineering teams at fast-growing hardware startups.

Let's think back to the C16 Bio case study in Chapter 4. One of their most difficult decisions was choosing the right initial market for their innovative synthetic palm oil product: Would they choose the large but resource-intensive food industry? Or the smaller yet innovative personal care industry? C16 Bio's founders eventually chose personal care, a market that allowed them to learn faster without scaling their production capabilities prematurely. These customers were also more open to new products and ideas. In other words, they were innovators.

In his 1962 book *Diffusion of Innovations*, author Everett Rogers developed a model for technology lifecycle adoption and defined five categories of customer profiles: innovators, early adopters, early majority, late majority, and laggards. You may be familiar with this concept from a more recent book: *Crossing the Chasm* by Geoffrey Moore.[23] Many startups make the mistake of targeting early or late majority customers, who are more pragmatic and therefore difficult to please. Instead, founders need to identify and target innovators who are willing to adopt an incomplete or buggy product because it solves (though not perfectly) such a painful need in their lives. Innovators and early adopters are also more willing to provide valuable feedback. AllSpice leaned heavily on their beta users when developing the early version of their product. Startups should build a solid foundation of loyal customers who can provide essential insights for further product development and market expansion, while also serving as enthusiastic advocates.

It is surprisingly easy to choose the wrong initial market and it's a mistake I see often. One founder I work with shared with me that it took her *six years* to figure out who her ideal initial customer really was. C16 Bio and AllSpice had numerous seemingly good options for their initial markets, but both eventually chose the best options at the time.

How do you ensure that you pick the right initial market? Here are a few criteria:

1. ***Pick an initial market that is consistent with your passion and mission.*** Focus your energy on a market and customer that you genuinely love. This will help you to get through the ups and downs, twists and turns of the early startup journey. One reason C16 Bio chose the personal care industry is that many companies cared deeply about the environmental impact of their products, aligning with the startup's core mission.

One way to find innovators is to look for the most passionate people in your space.

2. ***Select a market that is narrow enough to ensure focus, but large enough to attract capital and sustain multiple iterations.*** Entrepreneurs often pick too large or broad of an initial market. These markets are made up of majority-type customers who have high expectations and little patience for startups. On the other hand, some startups niche down too far and find themselves in a dead-end market. Your initial market should be a gateway to adjacent markets so that you can scale when you're ready. It's like the game *Frogger*. The goal is to hop across the river from one lily pad to the next. Your initial market is that first lily pad; if there are no other lily pads nearby, you'll be swept downstream. Make sure your initial market has a few lily pads nearby to continue the journey across the river. A good initial market size is one where you can build at least a $25–50 million dollar revenue business before needing to expand.

3. ***Identify a consistent set of features across customers.*** An initial market should not be so specialized that its features are not applicable to other markets. There is a place

in the world for customized software solutions, but not in Startupland (unless you're using AI to build customized solutions at scale). Be sure your initial market has a set of requirements that are common to a broad set of prospective customers. That way, it will be easier to dominate your initial segment and then jump to the next.

4. *Your initial market should demonstrate a high willingness to pay.* Many of AllSpice's freemium users were individuals who used Git, a free, open-sourced software version control tool. They were not accustomed nor interested in paying for a tool like AllSpice. This made them a poor choice for an initial market and why Ratner chose to target startup teams instead. Ask yourself: Does your initial market have a vibrant ecosystem of free products? Do your customers have the means to pay if you solve their problem?

5. *Ensure you have direct access to your customers (i.e., no gatekeepers).* Channels and partners are great for reaching more customers as you scale, but don't rely on these relationships to build your business in the early days. Choose an initial market where you have direct access to your customers so

you can learn from them. Staying close to early customers has been a key theme of all the startups we've explored so far: Squire bought their own barbershop, ClassPass worked directly with gym owners, C16 Bio created their own personal care brand, and AllSpice treated their beta users as design partners. Don't allow a gatekeeper to control the relationship with your initial market.

Your initial market should be full of innovators who desperately need your product. It's a jumping-off point to the next market, then the next, and then the next.

Growth Hypothesis: Attracting Customers

To quote the authors of Google's groundbreaking white paper on self-learning transformers (the invention that made gen AI possible): "Attention is all you need."[24]

Stories of innovative marketing, demand generation, and "growth hacking" have been part of Startupland lore for decades: Richard Branson's PR stunts, Hotmail's viral email signature, PayPal's referral program, Steve Jobs's legendary product announcements, Twitter's SXSW takeover, Duolingo's unhinged mascot, Steve Ballmer screaming the word "developers" fifteen times on stage... the list goes on.

These stories will pale in comparison to the growth "hacks" we'll see in the age of generative AI.

AI is already having a profound impact on the way companies gain attention—and their efficiency in doing so. Payments company Klarna is a good example of this. In their first half 2024 earnings report, they revealed the incredible extent to which AI has made them more efficient and profitable. Their average revenue per employee has grown 73 percent in the last year. Their customer service and operations costs are down 10 percent, and their sales and marketing costs are down 14 percent—all of which they attribute to their use of AI. "We can be more efficient in our segmentation, in our targeting, in our messaging . . . we have built AI copilots for each of the parts of the [marketing] workflow," said David Sandstrom, Klarna's Chief Marketing Officer, "I think that the best marketers are going to 10x their impact and efficiency because they have these tools."

Experimentation has always been a crucial part of startup growth strategy. You need to design a series of experiments, place small bets, and see what sticks. Then you cut the losing experiments and double down on the winning ones. AI now allows you to run more experiments faster than ever before.

Here is a simple example: A common growth experiment is to create a series of social media ads with different CVP messages to see which ones gain the most interest. Before generative AI, you might be able to create a dozen or so variations of the ad to run simultaneously and A/B test which ones are most impactful. Now you can generate hundreds and even thousands of variations at a time with a single click. The general LLMs like ChatGPT and Claude are already very good copywriters, but there are specialized tools being built for this distinct purpose as well.

Another example: SEO content marketing. In some markets, Google is the best channel for reaching customers. Think of markets where customers make many *high-intent* searches, meaning they are searching for immediate solutions (rather than idly browsing out of curiosity). One such example is the job search market. People looking for a new job regularly use Google and other search engines to find job openings, job descriptions, and sample résumés or cover letters.

David Fano, founder of Teal, had this insight in 2019 as he searched for an effective growth channel for his AI-powered job search platform. Teal builds a suite of tools for job seekers, including an AI-powered résumé builder and job application tracker to run a streamlined, organized job search. (Teal is also a Flybridge portfolio company, so I know their story well). Fano and his team experimented with several growth channels unsuccessfully, including social media ads and influencers. The problem was that most people saw these ads at a time when they weren't actively job searching. That's when Fano began to experiment with SEO. If he could establish Teal as a domain authority, Google and Bing would reward the startup with loads of high-intent search traffic. Fano hired content writers to create high-quality articles for job seekers, including job description and career path pages. Teal saw promising results from this channel early, but their growth was capped by the cost of content marketing. Fano was paying $500 per article.

Then ChatGPT came out, and Fano decided to go all in. He and his team built a low-touch AI writing system using OpenAI, no-code automation, and content management tools. Teal's GTM team created detailed article prompts and generated over three thousand articles tailored to specific job-related queries. For pennies, they created job-specific articles on career paths (e.g., "Prompt Engineer") and job-specific résumé templates (e.g., "CPG product manager"). Human editors reviewed each article and after minor edits and prompt tweaks, they were good to go.

The results were dramatic. Teal's organic search traffic grew to over 100,000 page views per week and converted 3 percent of all visitors into paid users, driving their CAC down to effectively zero. "The best content will win," said Fano, who believes job seekers don't care who, or what, wrote an article as long as it's useful.

Indeed, Google's AI-generative summaries prove exactly that: link clicks are down to historic lows due to these on-page answers from Google's own AI, Gemini. Still, Teal's SEO strategy is paying dividends. Many of their articles are cited by AI search answers, and high-intent job seekers still find their way to Teal's site to use its career growth tools.

Growth marketing is one of the fastest growing sectors for AI tools. Founders need to keep their finger on the pulse of new tools and experiment constantly. You never know when the next 10x innovation will be released. Even when you find a successful growth strategy, be on the lookout for the next one.

Sales Model Hypothesis: Closing the Deal

Attention is one thing. Closing deals is another.

There are three basic sales models in business. *Outbound Sales* is when a team of salespeople reach out to prospective customers to initiate the sales process. *Inbound Sales* is when the company attracts potential customers through marketing and promotion, then salespeople complete the sales process. *Product-Led Growth (PLG)* is when the company attracts potential customers through marketing and promotion, then the customer completes the sales process through a self-service portal. There are also distribution and wholesale sales models, but we'll touch on those in the next section.

In recent years, PLG has given startups a roadmap for rapid scale with limited resources. We have all encountered (and likely used) PLG tools in our work: Calendly, Slack, Canva, Figma, and others. PLG tools tend to be inherently viral by promoting teamwork, and they make it easy to sign up by offering always-free (dubbed "freemium") plans. PLG companies have mastered the art of SaaS: They remove virtually all friction and risk from the buying process. Still, it's not the perfect sales model for every company.

PLG works well for Teal, but not AllSpice, which uses an inbound sales model. What's the difference between these two startups? First, the customers. Teal attracts job seekers from virtually every field of work, but the intensity of their job search varies. Some folks are casually looking for their next thing while others treat the job search like, well, a job. For the power users, Teal offers a subscription with unlimited AI credits for résumé bullet points, professional summaries, and cover letters. Teal smartly offers this subscription on a weekly basis so job seekers can cancel as soon as they find a job.

AllSpice does not want casual users. During their beta period, they found that paying users were much more engaged with the product than freemium users. AllSpice aims to be part of the daily workflow for hardware engineers, so they want users to commit with their credit card. As a result, they canceled the free plans.

Another difference is product complexity. AllSpice's version control software is a technical tool for electrical engineers and their colleagues. AllSpice's freemium plan saw a lot of users sign up who were not good fits for the product. By talking with each new lead, the company can ensure AllSpice is the right tool for them. Teal is designed to be used by every job seeker from high schoolers to retirees. It's much more of a consumer-facing product, while AllSpice is a business-to-business tool.

Finally, the business models and unit economics of each startup are different. AllSpice has fewer customers to target (around 300,000 electrical engineers at 30,000 firms) and therefore needs a higher customer LTV to build a big business. AllSpice's plans start at a minimum of $64 per month per committer, and they only target teams of three or more. Their growth plan starts at $450 per month. Teal's total addressable market is in the tens, perhaps hundreds of millions of job seekers around the world, but their LTV is much smaller. They want to activate as many users as possible and then convert them once they prove their value, which their freemium plan helps them do.

We can again reference the ARPA sales curve to see why these two startups chose different sales models. Teal is attracting mice while AllSpice is targeting deer and elephants (and maybe one day, whales):

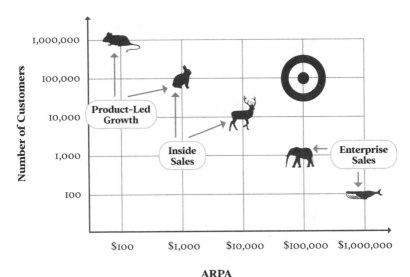

ARPA
(Average Revenue Per Account per year)

But as noted earlier, generative AI is starting to dramatically change this curve. To start, Teal and other PLG companies need fewer mice to be profitable and successful. AI has saved David Fano over $1 million a year on content creation alone. AllSpice is also growing more efficiently. They use AI to manage the majority of their inbound leads, leaving Ratner and her small sales team to focus on the elephants and whales.

Outbound sales have almost always been reserved for elephant and whale clients due to the high costs of fielding a sales team. But now AI has made the outbound sales model cost effective to attract smaller customers.

Topline Pro, a startup we discussed in Chapter 2, is a perfect example of this paradigm shift. Topline Pro creates custom websites and marketing content for service businesses. They charge between $250–$300 per month, putting their customers squarely in the rabbit range. In the recent past, the only way to cost-effectively reach this audience would have been PLG and *maybe* inbound sales. But Topline Pro founders Nick Ornitz and Shannon Kay had a problem: Service business professionals were an extremely fragmented market. There was no good way to attract those customers en masse, as is required with PLG and inbound sales. But with the launch of GPT 3 (the LLM model immediately before ChatGPT), Topline Pro had a brand-new opportunity: personalized outbound sales, powered by AI. Today, the startup reaches thousands of small service businesses every week with custom pitches—a feat that would have taken hundreds of sales reps before AI. Topline Pro is growing rapidly with just forty employees.

You should still be familiar with the traditional ARPA sales curve, but recognize you are no longer bound by time and cost restraints. You now have the freedom to choose the best sales model *for you and your initial market*, whether that's PLG, inbound sales, or outbound sales. The cost of sales is rapidly declining, regardless of the sales model.

But the best overall sales model is a mix of approaches. Yes, you can start with a single product tackling a single target customer through a single GTM tactic or motion. But over time, if a company is going to scale, they need to master multiple products for multiple customer segments through multiple GTM motions.

PLG is still an effective and efficient sales model for many startups early on. However, as PLG companies scale, they tend to become less efficient overall compared to their enterprise sales-led peers. This phenomenon is called the PLG trap.[25] Every company that aspires to sell to enterprise and secure "whale-like" relationships will have to eventually create an enterprise sales motion, a process that takes years to craft and perfect. If you rely solely on PLG or inbound sales to take a bottom-up approach through enterprises, you'll eventually be stuck. To avoid this trap, instrument the organization to operate a mix of sales motions after mastering the first one, including PLG, inbound, and outbound sales. Additionally, build the systems and expertise needed to service enterprise-level clients, including detailed and comprehensive documentation and a strong customer support program. Use AI across all these sales motions and post-sales functions to scale while staying lean.

Channel and Partner Choice Hypotheses: Teaming Up

Channels and partners (sometimes collectively called channel partners) are third-party companies or platforms that give you easy access to your target customer.

While they're often lumped together, there are key differences. Partners allow you to market and sell to your customers, while channels take on the responsibility of sales and marketing for you. Some channel partners are more possessive than others, meaning they keep control of the customer relationship.

Working with channel partners is a double-edged sword for both parties. There will always be a battle for customer loyalty. PayPal and eBay danced around each other for years as PayPal grew to become the dominant payment method on the much-larger eBay platform. eBay threatened numerous times to pull the plug on PayPal, which would have damaged the startup in its early days. Only PayPal's fierce customer loyalty among eBay users saved them.

Even if a channel partner doesn't aim to eliminate you, they will always try to charge higher and higher rents for access to their users. Facebook was a fantastic growth channel for numerous businesses from 2005–2012 until they throttled organic traffic to grow their paid ads business.

I always caution founders to be very selective in choosing channel partners—not only for the reasons stated above, but also because any layer between you and the customer will slow down learning. Again, you do not want gatekeepers to control your access to your customer base.

There are three questions to answer when exploring channel partners:

1. Is this partner the best way to reach customers?
2. Can we afford the incremental economic burden required to incent the channel partner?
3. How do we build an independent relationship with our customers that doesn't rely on the channel partner?

Let's look at how two very different startups sought to answer these questions.

Shippo: Riding Someone Else's Wave

Amazon has built a massive business controlling virtually every aspect of the buyer-seller relationship, charging enormous rents and forcing merchants to communicate to customers only through Amazon's dedicated portal. Shopify, the e-commerce platform on a mission to "arm the rebels" against Amazon, aims to be more benevolent to its partners and an open platform for third-party applications. Shipping startup Shippo found their footing by building the fulfillment backend infrastructure for small e-commerce companies in the Shopify ecosystem.

Inspired by the simplicity of Stripe's API for payments and Twilio's API for online communications, Shippo's co-founders, Laura Behrens Wu and Simon Kreuz, envisioned a similar tool for shipping that would streamline the process and reduce costs for online retailers. Shippo's initial product was an API that allowed businesses to integrate shipping functionalities into their systems, enabling them to compare rates, generate labels, track shipments, and handle returns seamlessly. Shippo's CVP opportunity was clear: "One single API connects our ecommerce stores to all the shipping APIs out there," explains Behrens Wu. "Our system is [like] the shipping backend of Amazon, but we make it available for small companies."[26]

However, selling the API was difficult. Small e-commerce stores didn't have the technical expertise, and midsize companies wouldn't trust a new startup with such a critical part of their business. So the co-founders went back to the drawing board. They decided to dogfood their own product by building a shipping app on top of their API. Shippo's app aggregated shipping rates from various carriers and provided an intuitive interface for merchants to compare rates, print labels, and track shipments. The app used a simple pay-as-you-go pricing model, charging $0.05 per label, which appealed to small businesses wary of long-term commitments.

Crucially, the co-founders listed their app in the Shopify marketplace, a new but rapidly growing channel for ecommerce tools. As one of the few shipping apps, Shippo gained traction quickly and earned an excellent reputation with their responsive customer support.

In just a few months, Shippo had 1,400 app customers, which in turn helped to build their credibility with larger API customers. Soon they were printing over 87,000 labels a month for app and API customers combined. The API customers were less numerous but did much higher volumes. Their churn rate was also virtually zero. As Shippo grew, they committed to supporting both customer segments.

Shippo rode the Shopify wave perfectly. Shopify had just 84,000 merchants on their platform when Shippo was founded in 2013. By 2024, Shopify had over 2 million. Shopify eventually invested in Shippo and made the startup their default shipping partner, solidifying their place on the channel.

It's still possible to grow a third-party application business through Shopify's marketplace today, but it will be much more difficult given its maturity. There are now thousands of apps, and hundreds in every category from shipping to design. Ideal channels change frequently, which is why you should never rely solely on a single channel for your success. Shippo smartly continued to build their own direct distribution even as their app took off within Shopify.

Ovia: Is There Product-Partner Fit?

Choosing a sales and distribution partner early in a venture's life is not as necessary as it used to be. It's simply easier than ever to reach a large number of customers through the Internet and online channels like Amazon, Facebook, and Shopify. But some industries still rely on partnerships more than others. If you're trying to disrupt an outdated industry, you're more likely to come across partners and middlemen who have been in the business for decades. You will have to decide to either play nice with them or innovate around them.

That was the question that Paris Wallace, founder of Ovia Health,[27] had to answer. Ovia Health was founded in 2012 to help people with fertility planning. The startup built an app for ovulation tracking, which later evolved into a comprehensive women's health monitoring platform. Ovia's early users were on the app constantly—recording their menstrual cycles, symptoms, and progress toward pregnancy.

After testing several business models (which we will detail in the next chapter), Ovia eventually caught the attention of large employers who wanted to provide women's health products in their benefits packages. But as with many enterprise sales motions in the pre-gen AI era, the sales cycles were long and expensive. Then Ovia was approached by a large health plan provider to become their exclusive channel partner. This deal would immediately bring in millions in revenue for the startup. How could they say no?

Wallace and his team looked deeply at this opportunity and found two issues: First, they would lose direct access to end users, which could severely hurt the startup long-term. Second, they would only earn about $0.10 per employee per month (PEPM), while the market rate for the app was closer to $0.50 PEPM. Was the scale of the distribution partnership worth it?

Wallace eventually turned down the deal and chose the slower but more sustainable path of selling directly to employers. This approach kept Ovia closer to their end user and allowed the startup to build a better product for both consumers and employers. After several years of learning and selling directly, Ovia eventually returned to health plan partnership deals. After a period of scaling substantial revenue and operations, the startup was acquired by LabCorp for a generous sum.

Wallace and the Ovia team made the wise decision to prioritize learning over rapid growth in the early days. They may have lost a multimillion dollar deal, but they built an enterprise that was many times more valuable.

Channel Partners and AI

Channel partnerships can be an enormous source of leverage, but they always come at a cost—both literal and in terms of customer connection. AI is not likely to play a massive role in building channel partner relationships, other than helping founders evaluate their strategic choices.

In fact, AI's biggest impact on channel partnerships may be their gradual demise. As it becomes easier to reach customers directly, as we saw in the case of Topline Pro, startups will rely less on third parties for their GTM; a good thing, in my opinion. While channel partners will never entirely go away, startups that own their access and relationship to customers will be more sustainable and profitable.

The principles of go-to-market strategy have not changed, but the means of reaching and acquiring customers are changing rapidly. As the customer acquisition costs go down, a startup's choices go up, opening doors to new, better, and magical business models.

Chapter 6

The 10x Club: Building a Magical Business Model

AT FLYBRIDGE, WE HEAR dozens of startup pitches every week. After each one, the partners and investment team debrief in private to debate some fundamental questions: If this company is successful, how valuable could it become? And crucially, does the startup have a "magical" business model?

AI is dramatically changing how we build startups, but the underlying fundamentals of a business remain the same. A business model still boils down to two things: how you make revenue (i.e., monetization) and how you earn profit (i.e., profit formula). AI can accelerate your search for a functioning business model and improve your unit economics, but the "shape" of highly valuable startups remains largely unchanged.

A magical business model is one that catapults a startup into the *10x Club*, that rare tier where a company is valued at over ten times its revenue.[28] There is a clear set of criteria to determine if a startup should be valued in the 10x Club; I will share those criteria later in the chapter. First, let's meet a 10x Club startup grappling with a crucial question: should you prioritize growth or profit?

Khatabook: Digital Transformation in the Heart of India's Retail Market

A significant portion of India's nearly trillion-dollar retail market comprises small merchants, including the more than 1.5 million family-owned convenience stores known locally as kirana shops. Many of those shopkeepers rely on traditional Bahi-khata paper ledgers to record their financial transactions—a process that is time-consuming and prone to errors. However, conditions in India are ripe for disruption. The proliferation of low-cost smartphones and data plans have rapidly increased Internet use nationwide. More kirana shop owners than ever are ready and able to digitize their businesses.

Serial entrepreneur Ravish Naresh discovered this opportunity in 2018 while participating in Y Combinator (YC) with a different company. He met the founders of India-based YC company OkCredit and was impressed with the core of their idea. "What stood out to me about their business model was that they were offering a substitute for cash transactions, which was still how most merchants and shopkeepers transacted," says Naresh. "Bringing off-line merchants online to transact with other merchants had an inherent viral flow. For every transaction, the app sent a notification to the purchaser alerting them of the transaction, even if they did not own the app."[29]

After he left YC and returned to India, Naresh couldn't stop thinking about OkCredit. The opportunity was massive—more than large enough for multiple players to capture significant portions of the rapidly digitizing customer base. Then Naresh discovered Khatabook, a mobile app developed and managed by a solo founder that tracked transactions and receivables for shopkeepers. The app was rudimentary yet still had 2,500 user reviews. OkCredit only had 1,500 reviews at the time. Clearly Khatabook was onto something.

Naresh quickly acquired Khatabook, kept the name, hired its founder to join his team, then rebuilt and relaunched the app. Naresh made Khatabook free, figuring that the key to success in such an enormous market was to drive massive adoption at all costs. The revamped Khatabook quickly gained traction among Indian shopkeepers. By February 2019, the number of new transacting merchants had jumped to 50,000; by April 2019, the app had 120,000 weekly users. This rapid growth secured the team a spot in Sequoia Surge, an accelerator program for Indian and Southeast Asian startups. With support and funding from this program, Khatabook was able to focus on improving post-install usage rate and expanding its user base.

As Khatabook's user base grew, so did investor confidence. In September 2019, the company closed a $26 million Series A round. By January 2020, Khatabook was adding a million plus new transacting users per month and showing promising retention metrics. Building on that momentum, the company raised a $60 million Series B in the spring of 2020.

But the startup's business model was a serious concern. Rather, their *lack* of business model is what worried investors most, even as they piled into Khatabook's Series B. While Khatabook's viral growth kept customer acquisition costs low, they hadn't run any monetization experiments and had no idea what the lifetime value of each customer would be. They had some hypotheses on how to make money—perhaps a premium subscription, a digital payments feature, or a credit line offering—but they had no actual data to reassure investors.

Luckily for Naresh, the lack of revenue was not enough to dissuade VCs. The startup financing market in 2021 was still frothy across the globe, driven by a combination of zero percent interest rates, a surge in digital adoption due to COVID spending, and investor enthusiasm for innovation. In August 2021, Khatabook was able to secure a $100 million Series C funding round at a $600 million valuation, punching their ticket into the 10x Club with multiples to spare. The message to Naresh from investors was clear: continue driving growth over monetization.

Khatabook raised just in the nick of time. As we covered in the AllSpice case study, the market turned soon afterward, and VCs shifted their guidance from growth to profit. As Bessemer Ventures put it in their *State of the Cloud 2023* report, "Unit economics [were] finally more in vogue... The market pullback ushers in a paradigm shift from the age of growth-at-all-costs to the age of efficiency."[30]

Like AllSpice, Khatabook got caught in the market shift. Naresh quickly recognized that the rules of the game had changed and began rapidly experimenting with business models. They tested several monetization strategies, including a lending-based model where Khatabook would facilitate small loans to merchants and capture the fees and interest as revenue. They also tried a B2B ecommerce marketplace model where they could capture transaction fees, as well as a freemium subscription plan. Khatabook also added advertising on their platform. On top of monetization, Naresh made the difficult choice to lay off dozens of employees to reduce their burn rate and narrow their losses.

Did Khatabook begin their monetization experiments too late, or right on time? We'll have to wait and see. As of this writing, the startup is still iterating on its business model. They had raised enough money in the grow-at-all-costs era to give themselves runway, but it is a race against time. As one Indian VC friend shared with me, "Viral growth led them to become one of the 'hot' companies in the market, but extremely overhyped." (I will address the tradeoffs between "blitzscaling" and incremental growth in Chapter 9.)

When to Monetize

To grow or to monetize? That is the question. Every startup will need to develop their business model sooner or later, but it's always a matter of timing. How do you know when you're ready to monetize? There are many factors to consider:

- *CVP and GTM fit:* In terms of sequencing, business model tests should follow your customer value proposition and go-to-market tests. Khatabook clearly discovered a hair-on-fire CVP and viral GTM plan. It was smart, in my opinion, to make their app free as they built their product and competed for market share. Not every startup can afford to give away their products for free—nor should they. However, wait to scale your monetization plan until you feel confident in your ability to attract and retain customers.

- *Market environment:* The cost of money is a major factor in how quickly you should start to monetize. In the zero interest rate era, plentiful capital was available and inexpensive; investors were eager to invest in promising founders even with few proof points of a sustainable business. Startups like AllSpice and Khatabook were encouraged to spend money to grow faster. If they could reach a certain scale and win their respective markets, there was a belief that startups could eventually be profitable over the long term. However, with higher interest rates and a higher cost of capital, startups need to be more conservative with their spend and try to monetize more quickly. If they run out of cash, they will have a much harder time attracting new investment. In my decades in Startupland, I have seen capital availability ebb and flow. Entrepreneurs need to be attuned to these shifts and calibrate their business plans accordingly.

- *Business model:* Some business models rely on much higher volumes of customers to be successful. For example, advertising business models need millions or billions of users. Enterprise SaaS companies require far fewer. The larger your customer base needs to be, the more it makes sense to prioritize growth over monetization.

At the end of the day, your decision to monetize comes down to this: Do you have enough cash to fund your operations and achieve your startup vision? Most startups take years to discover a valuable business model; that's why venture capital exists.

Understanding Startup Valuations

One popular definition of a startup, coined by Steve Blank, is "an organization formed to search for a repeatable and scalable business model." But you can't fund this search without money, which is why many startups take outside investment. It's crucial to understand how startup valuations are calculated, and how your business model hypotheses impact your ability to raise money.

At its core, a company's value is determined by its ability to generate future cash flows. Investors and acquirers use a variety of methods to estimate this value, but one of the most common is discounted cash flow analysis. This analysis projects a company's future cash flows and then discounts them back to their present value (recognizing that a dollar today is worth far more than a dollar in the future). It considers factors like growth rates, margins, and all kinds of risk to determine the extent of the discounting.

It's much easier to accurately value a mature, profit-generating company than a penniless startup. For example, Apple currently has a market capitalization of roughly $3 trillion. How did investors arrive at this number? They start by looking at the company's free cash flow, which amounts to roughly $100 billion a year. Then they account for factors on future cash flows: growth, competitors, and risk. They add up these discounted cash flows for the next several decades to arrive at $3 trillion. Said another way, investors believe Apple will generate an average of $100 billion in free cash flow every year for the next thirty years (for simplicity, assume no discount for the time value of money.)

The Saudi Arabian oil company Aramco also generates approximately $100 billion per year in free cash flow. But it is "only" valued at roughly $2 trillion. Why? Because investors have less confidence in its future cash flows compared to Apple. Investors are willing to pay more for Apple's future cash flows than Aramco's because they believe Apple's will be larger.

Valuing a startup is similar to valuing a mature company like Apple or Aramco, but with one key difference: Most startups don't generate free cash flow for years. In fact, many early-stage startups don't generate *any* cash flow at all. But investors will still try to value the company based on their beliefs about the future: *Will this startup, despite its negative (or lack of) cash flow, make a profit in the future? If so, how much? And how confident am I in this future coming to fruition?* All this uncertainty is why investing in startups is much riskier than investing in publicly traded companies. But it's a risk that VCs, founders, and founding teams take every day. In exchange, they have a chance to build the next billion- or trillion-dollar company.

Since startups don't actually generate free cash flow, VCs use *multiples of revenue* (or potential revenue) as a shorthand valuation tool. A startup could be worth 2x revenue, 5x revenue, 10x revenue, or even 100x revenue (at the time of its Series C, Khatabook was valued at 120x revenue).

Of course, valuations are used for more than just raising money. Most successful startup "exits" are in the form of an acquisition. Your eventual exit price is directly determined by your valuation. Even if you don't need to raise capital, you want to secure the highest-possible revenue multiplier you can.

Which brings us back to the 10x Club and "magical" business models. How do investors and acquirers know which multiple to apply to a startup?

Magical Business Models: Punching Your Ticket to the 10x Club

Not all business models are created equal. There is a reason some companies are valued at 10x their revenue while others are valued at only 1x revenue. That said, several common business model archetypes have proven successful for startups across a range of sectors.

One popular model is the *SaaS subscription model*. Companies like Cloudflare and Toast have built valuable businesses by offering their software as a service and charging customers a recurring subscription fee. This model provides predictable revenue streams, high customer lifetime value, and the potential for rapid growth through upselling and cross-selling. Because of these qualities, the SaaS business model routinely produces members of the 10x Club: Snowflake, Zoom, HubSpot, and Atlassian are just a few.

The consumer flavor of the SaaS subscription model is *consumer subscription*. Here, consumers are paying a monthly fee for a service. Spotify and Netflix are examples of popular consumer subscriptions that have been wildly successful. This model is generally considered less valuable than enterprise

SaaS because consumers are more fickle than enterprises and thus more likely to churn. For this reason, even the best of these types of companies are typically worth 6–8x revenue. In both SaaS and consumer subscription models, companies will often employ a *freemium sales model*. Think about Khatabook: the startup offered a free version of its product while it searched for monetization opportunities. Another example is MongoDB, a Flybridge portfolio company. MongoDB is now a publicly traded SaaS database company worth over $20 billion. MongoDB offers a free, open-source version of its product that provides some of the core functionality for customers who want to build personal projects. As soon as a customer wants more advanced features, they start a subscription payment plan.

Advertising is another tried-and-true business model, particularly for consumer-facing startups and media companies. Companies like YouTube and Facebook built large audiences and monetized those audiences through targeted ads. The key here is to balance monetization with user experience, ensuring that ads don't detract from the core value proposition. Because advertising-based revenue is more episodic than recurring revenue, these companies are typically less valuable—often even the best of them are worth just 3–5x revenue.

Another common archetype is the *marketplace* model, which has similar characteristics to the *transaction-based* model. With these models, companies build out platforms, attract users, and get a cut of the transaction volume on those platforms. Companies like Etsy connect buyers and sellers, taking a cut of each transaction. Financial technology companies (a.k.a., fintechs) like Block (formerly known as Square) or PayPal take a cut of every payment that runs through their network, much like older stalwarts like Visa and MasterCard. This model can be highly scalable and benefit from strong network effects as the platform grows. But their gross margins tend to be lower than SaaS companies' (e.g., Block's gross margin is 35 percent and PayPal's is 45 percent compared to Cloudflare's 80 percent). For this reason, the best of these companies are usually valued at 3–5x revenue.

Finally, there's the traditional model of *one-time product sales*, which is common in ecommerce. When Gap.com sells their latest jeans, they're making some margin on that purchase on a one-time basis. And, naturally, they're hoping happy customers will come back to buy more. This business model extends beyond physical products; a fintech that offers peer-to-peer loans is "selling" capital and then making money on the interest, but the revenue isn't recurring and the margins are relatively low. One-time product sale companies tend to be valued at 2–4x revenue.

As companies mature, they might blend multiple business models as they seek to broaden their product offerings. For example, Shopify charges merchants subscription fees to access their ecommerce platform (SaaS subscription), transaction fees for each sale made through the platform (transaction-based), and lending fees from their loan offering service (one-time sales). On the consumer side, Spotify offers a free version of its music service that is monetized with advertising. They also have a paid subscription that gives users ad-free listening along with offline capabilities and unlimited skips.

All else being equal, some business model archetypes are more likely to produce 10x Club startups than others:

Archetype	Revenue Multiples	Examples
SaaS	8 - 10x	Atlassian, Hubspot, MongoDB, Snowflake
Consumer subscription	6 - 8x	Life360, Netflix, Spotify
Advertising	3 - 5x	Facebook, Snap
Marketplace	3 - 5x	Booking.com, Etsy, Olo
Transaction-based	3 - 5x	Bill.com, Block, PayPal
Fintech/lending	3 - 5x	Affirm, SoFi
One time product sales	2 - 4x	Warby Parker, Wayfair

These revenue multiples are just guidelines. Within each archetype, there's a wide range in the actual multiples that companies achieve depending on their fundamentals, like growth and profitability. The business model is just a starting point. Afterward, investors evaluate each startup on its merits:[31]

- *A massive total available market (TAM):* TAM is the total potential revenue opportunity for your product or service. It's the size of the pie you're going after. But just because the pie is big doesn't mean you'll get a big slice. That's where the degree of product-market fit is considered, leading to confidence in a high market share or market penetration.

- *A strong competitive moat:* Popularized by legendary investor Warren Buffett, a competitive moat is your sustainable competitive advantage, protecting your product or service from rivals. If you have a strong competitive moat, you can retain that large market share or percentage of that massive TAM.

- *Recurring revenue:* The most attractive businesses are typically ones where customers are locked in and have high switching costs or high entanglement. For example, if all your favorite music playlists are stored on Spotify (I have dozens!), it is very difficult to leave Spotify for another music service. Plus, its personalized suggestions for new music are pretty good because it knows your history and the songs you like—this would be hard to replicate when switching to another music service.

- *Strong network effects:* No startup business model analysis is complete without discussing network effects. Network effects, popularized by technologist Bob Metcalfe, occur on platforms or products where each new user adds value to the existing user base. Uber is an obvious example—the more drivers there are, the more valuable the app is to riders, and the more riders there are, the more valuable Uber is to drivers. Even enterprise software companies can have network effects. For example, as more developers adopt MongoDB, it develops a richer ecosystem of tools and community support, making the platform even more appealing for new users.

- *Attractive unit economics or unit profitability:* This is where the rubber meets the road. For startups, the key metrics to consider are high gross margins (i.e., revenue minus costs of goods sold) and your LTV:CAC ratio—the lifetime value of a customer compared to the sales and marketing costs to acquire them. The higher your LTV:CAC, the more profitable your business is likely to be over time. How inherently profitable is your business at scale?

A company with these five components is very likely to earn its way into the 10x Club—regardless of their business model. A founder who exhibits command of these elements, even if their business doesn't check every one of these boxes, is going to inspire the confidence of investors.

Unit economics are particularly important to a startup's valuation; it's why I include LTV:CAC ratio as a core HUNCH metric for measuring product-market fit. But there are some misconceptions with LTV:CAC, which are crucial to explain.

Nailing Your Unit Economics

A former Flybridge colleague of mine left to become a managing director at a large, growth-stage investment firm. I caught up with him after a few months on the job and asked him, "What's different about how we taught you to think as an early-stage investor and how you now think as a late-stage investor?" His answer: An obsession with unit economics. We taught him the fundamentals of unit economics, but he didn't appreciate how important it was for founders to get it right to build a sustainable, enduring, valuable company. So let's take a minute to help you as a founder be a bit more obsessed with your unit economics.

A startup's unit economics primarily rests on its LTV:CAC ratio, which, as I noted before, should ideally be 3:1 or higher. The reason behind the 3:1 number is simple: An incremental customer should generate enough profit to not only cover acquisition costs but also operational costs and overhead. If you study the profit and loss statement of a company, there are three elements below the gross margin or gross profit line:

- *Sales & Marketing (S&M):* The cost of selling a product or service

- *Research & Development (R&D):* The cost of the engineering and product team

- *General & Administrative (G&A):* The cost of day-to-day operations and overhead (e.g., rent, utilities, executive salaries, finance, HR, legal)

In the LTV:CAC calculation, there is an allowance for the S&M costs (they factor into the CAC) but not the R&D and G&A costs. Thus, the extra 2:1 in the 3:1 formula is to cover these two cost elements.

Many startups calculate their LTV incorrectly, leading to flawed assumptions about their unit economics. I see three common mistakes entrepreneurs make when calculating LTV:

1. *Startups often forget to factor in gross margins:* I noted above that the three major cost buckets are below the gross margin line. Gross margins represent the money left over after the costs required to produce the product. For example, a software company's gross margins are often 70–80% after factoring in hosting, cloud, infrastructure, and data costs, as well as customer support

costs. Tesla's gross margins are 20 percent after factoring in the costs to manufacture the cars. If Tesla were to calculate their LTV based on their top-line revenue per sale (say, $100,000) instead of their gross margin (say, $20,000), it would be obviously wrong. Many startups either forget or perhaps overestimate their gross margins when projecting their unit economics. As a rule of thumb, I recommend discounting gross margin assumptions by an additional ten percentage points to be more conservative.

2. *Startups underestimate their churn rate:* Startups are often too bullish about how long they can hold on to their customers. They might extrapolate from early customer cohorts and assume customers will stick around and generate revenue for an unrealistically long time—I've seen some founders claim over ten years, which is hard to fathom in a competitive market environment. It's more realistic and appropriate to cap the expected months of revenue at three to five years (i.e., thirty-six to sixty months), depending on the nature of your customer relationship. As a benchmark, the average Netflix subscriber lifespan is thirty-three months, and the average Spotify subscriber lifespan is thirty months—and those are

two of the most popular and sticky services in the world!

3. ***Startups don't properly factor in their payback period:*** Imagine this: One startup brings in all its revenue immediately whereas the other brings in the same amount of revenue spread out over ten years. Which startup has a higher lifetime value? If you don't factor in the cost of capital, you would say they had the same value. That's obviously not correct. Startups have a very high cost of capital. They can't afford to support unprofitable customers for years and years in the hopes of earning a profit in a decade. As a result, shrewd founders will think carefully about their *payback period*—that is, how long it takes a newly acquired customer to generate enough income to cover their costs. My advice: Since a young public company's cost of capital is roughly 10–12 percent and a startup is considered much riskier, assume your cost of capital is around 30 percent per year. Therefore, future cash flows need to be heavily discounted when calculating LTV. A dollar of income earned one year from now is really only worth $0.70.

When you add up the impact of using conservative gross margins, more sensible time horizons, and a higher cost of capital, the resulting LTV can be dramatically lower than a naively-calculated figure—but more realistic.

It's crucial for startups to be rigorous and conservative in their LTV calculations to truly understand their unit economics. Savvy investors will see through inflated numbers. It is better to be truly conservative—or, dare I say, accurate—than to let a cynical investor do it for you.

In Search of a Functioning Business Model: Ovia Health

As of this writing, Khatabook is still in search for their ideal business model. I want to end this chapter instead with a case study of a startup that successfully found theirs. Let's return to the Ovia Health[32] story, our friends from Chapter 5, to see how they experimented their way to a winning business model.

Paris Wallace and his co-founders started the company inspired by the problem of helping people optimize their chances of conception. After achieving nascent product-market fit and growing their user base, they started experimenting with different business models. How would they make money?

Their first experiment was the freemium model—they'd allow users to download the app for free, then charge for premium functionality and content. They tested prices ranging from $1.99 to $9.99 per year. Conversion rates for the premium option were lower than anticipated, at times less than one percent. The team decided to abandon that experiment not only because of its poor performance, but because it felt misaligned with the company's vision. Customers benefited from having more customers in the community; paywalling created friction in building their network effects, which not only hurt their customers but also their long-term valuation.

The Ovia team then embarked on their next business model experiment: advertising. Users often asked for recommendations for baby and fertility products. With such an active and motivated user base, Wallace thought they could charge advertisers a premium. An MVP advertising network was quickly developed and Wallace led sales. Despite having no experience in advertising, he was able to earn over $1 million in advertising sales in the first year.

Advertising seemed promising, but Wallace calculated the startup's potential valuation with an ad-based business model. His math: Four million pregnant women each year, plus four million women trying to conceive, meant a total possible audience of eight million users a year. At an advertising value of $12 per user, that meant the TAM would be roughly $100 million in annual revenue. This was far too small of a market to match the team's ambitions. Further, Wallace knew that ad-based business models were typically valued at lower multiples than subscription models, thus hurting their chances of joining the 10x Club.

The team then launched on a third monetization experiment, and this one was the winner. Wallace started talking to large employers and health plans about their platform and user data. He learned insurers wanted to identify pregnant people, and especially high-risk pregnancies, so that they could enroll their members into preventive wellness programs. Not only that, but insurers and employers were willing to pay a hefty annual subscription to access this data. After a successful pilot program with one large insurer, Wallace built a sales team and expanded the enterprise SaaS model to dozens of large customers. He also discovered an unexpected acquisition strategy: Ovia users were willing to share their insurance provider in the app, which gave Wallace's team a strong proof point in sales conversations.

Ovia Health was eventually acquired by Labcorp for over $100 million. At the time of the acquisition, it was reported that their valuation was north of 10x revenue. Ovia is a great example of a company that experimented and iterated its way to a successful profit formula, maximizing its chance of building a truly valuable company. It's not easy, but for those who get it right, the rewards can be immense.

So now that we've covered what makes a magical business model, it's time to scale.

In the next three chapters, we'll explore how AI is transforming the way startups build, grow, and operate. We'll look at integrating AI into product development, operations, supply chain management, and more. We'll also examine the challenges of becoming an AI-forward company, and how startups can position themselves to thrive in the age of intelligent automation.

Chapter 7

Scaling Product: Growing Beyond Product-Market Fit

THE FIRST STARTUP I joined after graduating from business school was an Internet 1.0 rocket ship called Open Market. We ran around in circles chasing product-market fit for a year or two, and then hit it big time. Our revenue grew from $1 million to over $20 million in one year, and we were in the process of tripling the next year when we decided to file for IPO on the NASDAQ exchange. On our first day of trading, we achieved unicorn status—a mere three years after our founding, we were worth more than $1 billion. In the mid-1990s, that was very, very rare.

At the IPO party, my mentor and boss, the CEO, took me aside and confided in me. "Things are about to get much harder," he warned. "It's like the story of the dog that chases the bus."

"What do you mean?" I responded.

"We caught the bus," he said, smiling. "What on earth are we going to do with it?"

What I understand now—but didn't fully appreciate then—is that finding *strong* or *extreme* product-market fit is just the beginning. Scaling your startup is even harder.

What is Scaling?

Scaling is the process of growing a startup's operations, revenue, and impact in a sustainable and efficient way. Companies often stumble at this stage because they don't have the knowledge, tools, or team to build on their initial success. Getting *scale* right is critical to building an enduring, valuable company; getting it wrong could mean years of floundering as you burn through cash and lose your advantage. Many startups never recover after failing to scale.

The challenge is that scaling often requires you to build a completely new business while running the old one. The degree of difficulty and operational complexity ratchets up dramatically. Multiple products are required to add to the "layer cake" of the business—and those multiple products might serve multiple customers, and those customers might be spread out across disparate geographies and reached through multiple channels. The talented team that got you to this point may not be the right team to lead through hypergrowth.

As if it weren't hard enough from a purely logistical standpoint, the expectations surrounding a rapid-growth company can add crippling pressure for a founder. At this point, as we experienced at Open Market, you may have crossed a few million dollars in revenue and find yourself on the coveted "hockey stick" growth curve. Now your team starts doing what I call phantom equity math ("If this company were worth a billion dollars, my shares will be worth X, I'll be a multimillionaire!"), and your early investors move you from the "Too Early to Tell" column to the "High Return Potential" column on their portfolio Kanban boards.

Artificial intelligence adds another wrinkle to the challenge of scaling. It unlocks startups from the shackles of capital and human headcount to grow faster and leaner than ever. But it's also a powerful and sometimes unwieldy tool that founders must learn to harness.

Scaling requires a delicate orchestration of people, resources, and technology. Many years ago, I was in the middle of scaling my second startup and was featured in a Harvard Magazine article on entrepreneurship. In it, I described the founder's job in a growth-stage startup as "the dance of the dreamweaver."

Founders sell VCs on the size of the opportunity ("Trust me, this is going to be huuuge") and promise them a cadre of seasoned managers to bring it to life. Founders then recruit those senior managers by assuring them of adequate financing and strong demand. All the while, the founder is attempting to land customers and partners with assurances of financial strength and managerial aptitude. Having set this cycle of spiraling expectations into motion, when you get back home, you have to sweat like hell to make it all come true.[33]

This is what it feels like to scale, and why the world of VC-backed startups is not for everyone. But for the brave, it's a thrill like none other.

Before we get into the mechanics of scaling product, let's answer an important question: When?

When is it Time to Scale?

Paul Graham's advice to "do things that don't scale" in the early days is great. Until it is time to scale.

Before embarking on a scaling journey, startups must carefully evaluate their readiness. There are many famous startup failure stories that resulted from premature scaling. My colleague, Professor Tom Eisenmann, refers to these signals as "false positives" in his book on startup failure.[34] Eisenmann points out that founders are especially susceptible to false positives because they are "psychologically wired to see what [they] want and hope to see."

How can you avoid falling into the false positives trap? First, test your idea with mainstream customers. If they don't get it the way your early adopters did, you may find yourself stuck in Geoffrey Moore's "chasm"—the gap in customer expectations between early adopters and early majority users.

Eisenmann lays out something known as the RAWI Test for entrepreneurs to help them assess whether they are prepared to push the scale button. The test asks four questions:

- *Ready?* Are you seeing strong demand for the product, and are you satisfying customers?

- *Able?* Do you have access to the necessary human, capital, and technology resources required to scale successfully—that is, is the team in place and can capital be raised?

- **Willing?** Are you committed to growing the business, despite the risks of scaling, and will growth advance your original vision?

- **Impelled?** Does the startup have aggressive rivals and is the market characterized by strong competitive intensity? Is speed and scale imperative to success?

These questions offer guidance for founders, but you can also take a quantitative approach to measuring your readiness to scale. Specifically, I recommend grading your startup across the HUNCH metrics and the strength of your product-market fit:

	Nascent	Developing	Strong	Extreme
Hair on Fire CVP (40% test)	20%	30%	40%	50%
Usage High	Varies by company and industry			
Net Promoter Score	40	50	60	70
Churn Low	B2C: 8%/Month B2B: 5%/Month	6%/Month 3%/Month	4%/Month 1%/Month	2%/Month Neg. Churn
High LTV: CAC	2:1	3:1	4:1	5:1

HUNCH Benchmarks at each stage of Product-Market Fit

Most startups should wait until they see indications of *strong* product-market fit before investing time and resources into rapid growth:

- 40 percent or better on the hair-on-fire test
- High usage rates (which vary by industry)
- A net promoter score greater than 60
- Low churn (less than 4 percent/month for B2C and 1 percent for B2B)
- An LTV:CAC of greater than 3:1 (ideally closer to 4:1)

But knowing when to scale is ultimately a judgment call. It's similar to the question of when to monetize; it depends just as much on the market environment as it does with what's going on inside the business.

If the RAWI test and HUNCH metrics indicate to you that it's time to scale, then you will embark on the dance of the dreamweaver: building one aspect of the company while securing resources for another, all the while servicing existing customers and ensuring that the house of cards doesn't all come crashing down.

Scaling AI-Forward Products

The entire premise of the 10x Founder is that founders and founding teams can scale their organizations *without* growing headcount as rapidly as startups had to in the past. The most expensive line item for most startups is their technical talent:

the engineers, developers, and product people who bring technology products to life. With AI copilots and agents, startups can now build well beyond an MVP-level prototype to scale their product without amassing an enormous engineering budget. "Software engineering sees the greatest benefit [from generative AI]," said VC Tomasz Tunguz in a 2023 blog post.[35] "Comprising 40% of the company and producing 25% more output (GitHub says 50% of all code is now machine-generated), the startup should see a 10% overall [financial] improvement."

First let's talk about AI copilots, the "mature" technology in the space, and then discuss the cutting edge (as of this writing) with agents.

AI Copilots: Super Clippy

The gen AI revolution has been good for the legacy of Clippy, the former mascot of Microsoft Office that bounced around in the corner of our desktops for over a decade. Clippy was a paperclip with eyes that helped users navigate the impenetrable MS Word interface. It could tell you how to change your paragraph settings and would remind you to save your documents.[36] Beyond that, it was about as useful as an actual paperclip. But now that gen AI has begun to deliver true copilot functionality, many folks are speculating when Microsoft will revive Clippy and give it superpowers with OpenAI's latest GPT.

Regardless of Microsoft's plans for Word, we all now have "Super Clippys" at our fingertips across virtually every application: word processing, image generation, spreadsheets, and particularly code development. LLMs excel at producing and reproducing language, including computer languages. Companies large and small are unlocking new levels of productivity thanks to coding copilots like GitHub Copilot, GitLab Duo, and Cursor. The general LLMs like ChatGPT, Claude, and Llama are also extremely capable code generation copilots. Not only are non-technical people able to code up impressive projects (like my founder friend who replaced their CTO with ChatGPT), these tools also turn 10x developers into 100x developers.

The impact of AI copilots has already been enormous, and we're just getting started. In October 2024, Google shared that 25 percent of all new code within the company is produced by AI.[37] Let me repeat: one out of every four lines of code at Google is AI-generated. And as these tools get better, they will only become more prolific and efficient.

Most MVP-stage products are not capable of handling the demands of scale. Expanding, or even rebuilding the underlying technology is often necessary. For more expansive development capabilities, we need to turn our gaze to the future of Agentic AI.

Agentic AI: The Next Frontier is Here

Studies have found that AI copilots make engineers about 26 percent more productive.[38] AI agents have the power to make us *100 times* more productive.

Agentic AI represents the shift from copilot to full decisioning and action-taking. We will have general use AI agents for our day-to-day lives (e.g., "Call me an Uber to take me to the airport") and also specialized agents designed to complete specific tasks, series of tasks, or to work on teams with other AI agents.

Agent teams have the greatest potential impact on the way startups scale. Imagine a team of specialized agents working together to move a software project from ideation, through development, through testing, and into production. Hard to imagine? This is exactly what startup Blitzy.AI is doing today to build enterprise software in one-tenth the time.

"Software development today is AI-supported, where you have engineers using copilots," said Siddhant Pardeshi, co-founder and CTO of Blitzy. "But what if you could flip that and have AI do most of the work, and only have the humans fill in and do the rest?"

Blitzy's vision is a world where *all* B2B software is custom-built to the exact specifications of the customer. This world is not as far off as you might think. Today, 80 percent of all commercial software is built on open-source technology. That means AI has seen and ingested it all as training data. AI models have already learned how to build good software, they just need the *agency* to do the job.

That's what Blitzy is trying to do. The startup recently won a contract for an enterprise software project that required 30,000 lines of code and would normally take six months to build. They completed the project in just six days with one senior engineer (their next goal is to complete projects like this with a junior engineer).

Their process is cutting-edge, but it's a harbinger of things soon to be available to the masses. The key is to think about breaking down complex tasks into smaller, specialized steps that can be handled by different agents while maintaining human oversight for quality control and final refinement. Here is an overview of how Blitzy's AI agent system builds software:

Step 1: A specialized agent takes the initial product vision and concept, and translates it into a product requirements document (PRD). This is the roadmap for all subsequent development.

Step 2: A second agent ingests the PRD and expands it into detailed technical requirements.

Step 3: A third agent converts the technical requirements into technical specifications. This is the step that senior engineers would traditionally take on. The specs are then reviewed and finalized by human engineers.

Step 4: A fourth agent translates specifications into system architecture. It maps out how different components of the software will work together and handles complex system dependencies and integrations.

Step 5: Finally, thousands of coding agents (up to 3,000) work collaboratively to build the software. Each agent handles a specific portion of the codebase. Together they can generate 30,000–50,000 lines of code in just a few hours. This is how Blitzy overcomes the token limit of individual AI models (which, as of this writing, is 8,000 tokens or approximately 6,000 words for ChatGPT).

The entire development process currently takes about twelve hours. Ten of those hours are spent on "thinking" through the first four steps. Only the final two hours are spent actually coding. Right now, Blitzy promises to create about 80 percent of a final project, leaving the final 20 percent for human engineers.

Blitzy is not alone; Replit and Codeium are startups with similar platforms growing in popularity. We are still in the early innings of the Agentic Revolution.

On LinkedIn, someone asked in the comments of one of my posts, "If AI can build software all on its own, why even have human founders?" The answer is simple: The jobs of strategy and leadership will remain in human hands for the foreseeable future. AI can now build software, but it can't decide for you *what* to build or how to create a business around it. Most importantly, AI can't set your company's culture; only you (the founders and leaders) can do that. "Leadership has always been done the old-fashioned way," noted Brian Elliott, Blitzy co-founder and CEO in a conversation we had. "If you want your company to be AI-native, you have to lead and be AI-native yourself and say, 'How can we solve this problem with AI as a starting point?'"

Scaling your product in the age of AI is not only about utilizing AI copilots and agents to accelerate your work, but building a culture of AI-first problem-solving and experimentation.

Managing Tech Debt

As startups grow, their tech infrastructure must keep pace. This means moving from basic systems to more robust platforms, which can prove to be a challenge if you didn't properly plan for scale. Let's use our friends at Shippo, the shipping and logistics startup we discussed in Chapter 6, as an example.

Laura Behrens Wu and Simon Kreuz got the idea for Shippo after encountering the high costs and complexities of shipping handmade goods internationally. That led to Shippo's initial product: an API designed to streamline business shipping processes. The value proposition for their API was clear—the software would enable businesses to compare shipping rates, generate shipping labels, track shipments, and handle returns on a single, integrated platform. Early on, however, Behrens Wu and Kreuz struggled to land midsize and enterprise customers, who were skeptical of relying on an unproven solution. This forced the team to build a Shopify app designed for small businesses who lacked the technical chops to integrate an API into their systems. The pivot worked; it gave Shippo the experience and reputation needed to win over bigger customers. Eventually they doubled down on the API and committed to selling upmarket to larger customers.

But they made one mistake. Instead of building the Shippo dashboard on top of their own API, they built the app *alongside* it. That meant the app would sometimes bypass the API entirely to access the startup's core functions and services. On one hand, the app and API could evolve independently. On the other, they had to maintain and improve two separate code bases, creating large engineering inefficiencies. "When we added a new API functionality, we had to add a UI [user interface] element to the app to allow the user to interact with it," said Behrens Wu.[39]

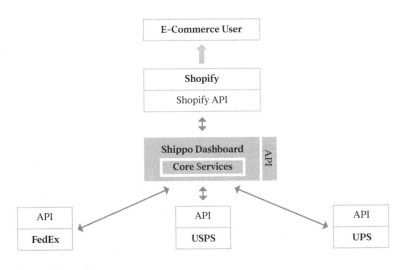

Source: Casewriter

Shippo's App and API Development Strategy

> *Notice how the Shippo dashboard and API were built alongside each other rather than the app on top of the API.*

As Shippo shifted their focus back to the API, they had to spend countless hours *rewriting* and refactoring their entire code base. This oversight almost took down Shippo, and very well could have wiped out a startup with less money in the bank. As liberal arts majors (Kreuz, the CTO, later taught himself how to code), the co-founders underestimated the costs of *technical debt*, the very real drag on resources that comes from having to rewrite and recreate key parts of your codebase.

Luckily this story has a happy ending. Shippo survived its technical debt crunch and rebuilt its API. They also maintained and improved their Shopify app and eventually became the official shipping logistics platform for companies of all sizes. By 2023, Shippo had over 100,000 total business customers. Says Behrens Wu, "The first iteration of the app was maybe built in a day or two. And we're still dealing with that sort of tech debt, but it got us to market pretty fast."[40]

Accruing some level of technical debt is expected, but not all tech debt is created equal. You can think of some types as "low-interest" debt that are easy to pay off and others as "high-interest" that could end up costing you more than the original development. Software developer Martin Fowler, author of *Refactoring: Improving the Design of Existing Code*, posits that there are four kinds of technical debt:[41]

1. ***Deliberate and reckless debt*** happens when a team knowingly cuts corners to get quick results, which often causes problems later for startups.

2. *Deliberate and prudent debt* occurs when a team intentionally pushes off some work for strategic reasons and plans to fix it later.

3. *Inadvertent and reckless debt* comes from not knowing good design practices, leading to messy code that slows progress.

4. *Inadvertent and prudent debt* is the natural result of learning during a project. Even experienced teams realize they need to improve their initial design as they go along. This type of debt is typical and expected as the team gains more understanding of how to solve the problem and achieve the objective of the code over time.

AI copilots and agents will mostly eliminate inadvertent and reckless debt. There is no longer any excuse for startups to use poor design practices, even for non-technical founders like Kreuz and Behrens Wu. AI copilots will guide you through best practices and AI agents will build on them. We might see less tech debt when AI agents can improve both development speed and quality by 100x. But then again, any startup truly pushing the cutting edge of what's possible will accrue technical debt as they slice their way through the unpaved jungle. In these cases, AI coding agents will be valuable tools for refactoring and rewriting code bases in a fraction of the time.

Scaling product is always a balance between anticipating demand and conserving engineering resources. AI is making this balance a little easier to manage; it now requires fewer resources to build the robust infrastructure needed as you grow. Now let's talk about demand: scaling go-to-market.

Chapter 8

Scaling Go-To-Market: Sales, Marketing and Expanding TAM

AFTER ENGINEERING, GO-TO-MARKET IS the largest area for AI-powered productivity gains within high-growth startups. It's also riddled with pitfalls like the PLG trap (discussed in Chapter 5) and the lure of prestige hires.

Sales is particularly challenging to scale because it looks so different at different stages of a business. It's surprisingly common for startups to hire the wrong sales leader for their stage of growth, often leading to disaster. Properly scaling sales starts with understanding the Sales Learning Curve.

The Sales Learning Curve

The Sales Learning Curve, a concept introduced by Mark Leslie and Charles Holloway in a 2006 *Harvard Business Review* article,[42] is a crucial framework for startups transitioning from founder-led sales to a scalable sales organization. I have taken this general concept and applied it more specifically to startups in my HBS course, Launching Tech Ventures.

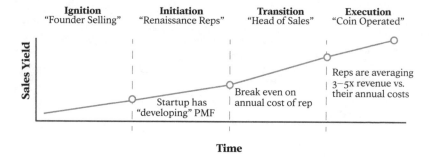

Sales Learning Curve

AI sales tools, like those we covered in Chapter 5, are compressing this curve horizontally and vertically. Startup sales teams will be able to accomplish more with less while accelerating their sales yield per employee. Venture capitalist Tomasz Tunguz predicts at least a 15 percent gain in productivity with *today's* tools;[43] that will only improve as the models and tools built on top of them get better.

Keep this in mind as we review the stages of the sales learning curve below. The basic structure will stay the same, but your journey through the curve will accelerate as new tools come online.

Stage 1: Founder Selling

In the earliest days of your business, sales should be led by a founder. There is no one on the planet with as much passion or insight into the specific problem you're trying to solve. Plus, early sales should be viewed as a learning exercise as much as a commercial one. Hands-on sales experience gives founders the intimate understanding of customer needs, objections, and decision-making processes needed to scale later on. The goal at this stage is to uncover customer insights to improve product-market fit.

AI gives founder-sellers a massive advantage by saving them time on manual prospecting, customer research, and outreach. My colleagues and I at Flybridge are tracking over sixty startups that are building new AI-powered capabilities in this space alone, and a few of them—most notably Clay and 11x—are gaining substantial early traction. The specific tools are less important given how rapidly the market is evolving. More important is the way you use these tools; don't let technology get in the way of face-to-face conversations. Think of AI as leverage that frees you up to have *more* conversations with customers, not fewer.

How long should a founder personally lead sales? While there's no hard-and-fast rule, founders should try to lead sales at least until they have *developing* product-market fit. In general, AI has allowed founders to lead sales for longer than ever, saving the startup money and keeping a close connection with the customer.

Stage 2: Renaissance Reps

A startup will eventually hire a salesperson. But who do they hire? A big mistake I see often is an eager founder who hires the most experienced sales rep they can find. This person comes with an impressive résumé full of enterprise sales experience at blue chip companies, and they exude professionalism and gravitas. This is exactly the *wrong* person to hire.

The job of your first salesperson is not to scale your startup to a billion dollars in ARR; it's to help you *build* your sales process. I call these folks *renaissance reps*. They can really do it all: prospect and sell, but also create systems and pass on learnings to the product team. Founders should think of these hires as business development co-founders; they need to be curious, creative, adaptable, and willing to roll up their sleeves when needed. Besides those traits, though, renaissance reps come in different stripes. Some have technical backgrounds and others will come from the customer side. Hire one of each type of rep to see which one has more success.

It's critical that the founder-seller does not abdicate responsibility at this stage. Their role shifts from sales rep to sales leader. Their job is to coach their renaissance reps and systematize the sales process. This job includes:

- Hiring, onboarding, and ramping up an initial batch of sales reps to achieve consistent sales results

- Establishing and documenting a sales process that new reps can follow

- Developing a reliable system for tracking key sales metrics and performance

- Creating a set of standardized sales tools and sales training materials that ensure a consistent message and approach (an area where AI can really help)

You know you're ready to move on to Stage 3 of the sales learning curve when you break even on the cost of your sales reps. But again, AI is compressing and warping the curve. You may reach this breakeven in half the time you would have before. Founders can either continue scaling sales in an effort to accelerate, or maintain their fleet of renaissance reps to save money and stay closer to the customer.

Stage 3: First Head of Sales

The founder can't lead sales forever (though they can lead it for much longer than before). After finding success with a small team of renaissance reps, a founder should consider hiring a head of sales. Again, many founders mistakenly hire the most senior, pedigreed sales leader they can land. Avoid this temptation at all costs. At this stage of growth, you need a "player coach"—someone who is able and willing to sell directly but can also manage and coach up their reps. This requires a leader skilled in managing people and systems, capable of overseeing dozens of sales representatives, all the while maintaining the scrappiness established in earlier stages.

The mission of the first head of sales is to make improvements to whatever sales system the founder-seller set up in the first two stages of the Sales Learning Curve. This could include:

- Upgrading technology, including CRM, customer discovery, and AI-powered outreach

- Refining the customer segment for more effective prospecting

- Improving sales scripts to improve the team's close rate

- Offering higher-level coaching to reps to improve individual performance

- Shifting the team from a relationship-driven sales culture to a data-driven, performance-based culture

A final qualification needed for your first head of sales is their level of comfort with AI. When used properly, AI GTM tools will allow your team to hit their targets faster and with fewer sales reps.

The revenue target for this stage of growth is 3–5x revenue on the annual, fully loaded costs of your sales team. For example, if an individual rep might make $200,000 at their on-target earnings level, then their sales quota should be $600,000 to $1 million.

In the final stage, your sales team and culture shifts from a small, adaptable team to "coin-operated" selling machines.

Stage 4: Coin-Operated Selling

Once a sales leader is achieving consistent growth and has proven their ability to hire and onboard new sales reps, they must evolve from building a sales team to scaling and optimizing a sales *organization*. This stage is called "coin-operated" selling: sales reps are highly motivated by financial rewards to close deals and hit their numbers, and they do so repeatably and predictably. Coins in, sales results out—like a well-oiled machine.

The coin-operated selling stage looks completely different from the founder-led and renaissance reps stages. Once the machine has been built, you can begin to hire more traditional sales leaders and reps with enterprise experience. Quotas are handed out and made, forecasts are relatively accurate, marketing and sales are aligned and working harmoniously, and the sales team can be counted on to deliver results quarter after quarter.[44]

I will say again: It is critical you don't try to skip straight to Stage 4, even after finding early success in the founder-selling stage. Even with the productivity gains of AI, young startups do not have the culture or infrastructure to run like a coin-operated machine. Most of all, they don't have the customer knowledge or extreme product-market fit needed to be successful.

To illustrate this point even further, let's revisit the Ovia Health story. Initially, Ovia's co-founder and CEO, Paris Wallace, found success with founder-led sales (Stage 1), particularly in selling to employers and health plans. But then, in an effort to grow quickly, Wallace attempted the leap to Stage 4: coin-operated selling. He hired an experienced VP of sales with deep expertise in the HR and health insurance space, plus six experienced sales reps. All of these folks had impressive résumés and knew the industry better than Wallace—but none of that mattered. Ovia didn't have a mature sales playbook or system, and Wallace's new superstar salespeople weren't the types to build them. Productivity stunk, and Wallace had to fire the entire team after just six months. He rebooted with less-experienced but scrappy *renaissance reps* and took back the role as head of sales. Wallace and his new team built the playbook from scratch and learned how to sell first-hand. Wallace eventually hired a "player-coach" type head of sales, who continued to refine and optimize the program. Ovia eventually reached $20 million ARR and was acquired in an all-cash deal.

Once your startup sees success, you will be flooded with opportunities to hire experienced sales reps and sales leaders who are eager for a big payday. Don't be tempted by prestige hires.

Scaling Marketing: Content and Experiments

During the development of early product-market fit, a successful startup will discover one, maybe two effective growth channels where they can reach and convert their target customers for relatively cheap. The initial channels could be SEO content marketing (like Teal), community-driven marketing (AllSpice), product-led growth (Khatabook), or paid advertising. But in the game of customer acquisition, good things never last. Your first channels will eventually become less effective as you convert all the convertible customers, or they get too expensive as platforms engage in rent-seeking (e.g., Google suppressing organic links to drive paid advertising). So the question is, what do you do when your initial growth channel dries up? Experiment, experiment, experiment!

Marketing often remains the most creative and scrappy team inside a growing startup because "what works" is constantly changing. Marketers must always be on the search for the next winning growth channel. Think of it like oil prospecting: once you discover a gusher, your job is to go find the next one. Luckily, your marketing team may *feel* the biggest impact of AI on their work. Tomasz Tunguz estimates that marketing teams will experience up to 30 percent productivity gains thanks to current AI tools (but this only adds up to 1.5 percent in total organizational productivity due to the relative size and budget of marketing compared to engineering and sales).[45]

The advantage of AI for marketers is scale. Instead of testing two or three versions of ad copy or landing pages, startups can now generate and test dozens of variations simultaneously. One of my portfolio companies recently used ChatGPT to generate fifty different versions of their value proposition to test with different customer segments. They discovered messaging that resonated three times better than their original copy—an insight that would have taken weeks to uncover manually.

AI also enables *personalization* at scale. Think back to Topline Pro, whom we met in Chapters 2 and 5. The startup creates thousands of personalized pitches for home service businesses with custom-built AI tools. Their cold email response rates jumped from zero to nearly 10 percent with a fraction of the staff traditionally needed.

Content marketing remains one of the most capital-efficient channels, but it's historically been time and resource intensive. We saw in Chapter 5 how David Fano at Teal used AI to generate thousands of SEO-optimized articles, dramatically reducing their customer acquisition costs. The key was having clear editorial guidelines and human oversight of the AI-generated content. Their organic search traffic grew to over 100,000 page views per week and converted 3 percent of all visitors into paid users.

However, AI is not a silver bullet. It helps you run experiments faster, but it doesn't tell you which experiments to run. That still requires strategic thinking and solid marketing fundamentals.

PLG and Avoiding the Trap

Product-led growth (PLG) has become the default go-to-market strategy for many startups, and for good reason. When it works, PLG is magical—users discover, try, and buy your product without ever talking to a sales rep. Dropbox, Slack, and Zoom all rode PLG strategies to billions in revenue. Even enterprise software companies like MongoDB and Snowflake have embraced PLG to accelerate their growth.

But PLG is not a panacea. As we covered in Chapter 5, startups relying on PLG to move bottom-up into enterprises can get capped by their lack of enterprise experience. This is called the PLG Trap. Inspired by their success with end users, a PLG company may try to expand their outbound sales team rapidly and sell directly to executives. They quickly learn that the needs of end users are not the same as those at the top of the organization. PLG companies typically don't have the customer support and sales engineering teams required to service enterprises. If a startup reaches this point, they will be stuck with a massively expensive outbound sales team and no results to show for it.

How do you avoid the PLG trap? The answer is to embrace PLG but plan ahead. Building an enterprise sales motion is a years-long process. If you have found success with PLG, start experimenting with inbound and outbound sales *today* to build the muscle. AllSpice is a good example of a startup doing this. While they aren't purely PLG, they have a strong inbound marketing program that drives the majority of their sales leads. Co-founder and CEO Valentina Ratner has led the startup's sales team since day one despite having no sales experience herself. She is learning first-hand what it takes to sell directly into small businesses as well as enterprises. As AllSpice grows, I expect their enterprise sales motion to be ready to grow with them.

The PLG trap is especially dangerous in the age of AI, where products can be technically impressive but still fail to gain enterprise adoption. Just because individual users love your AI product doesn't mean enterprises will trust it with mission-critical workflows. The most successful startups blend PLG with other go-to-market motions. Product-led growth creates initial user love and bottom-up adoption; sales-led growth builds the enterprise relationships needed for large contracts; and inbound marketing provides air cover, education, and demand generation across both motions. Deliver product value across multiple levels of the organization from end user to C-suite.

Expanding TAM: Scaling to New Markets

The ideal initial market for a startup is one that can support a $25–50 million revenue business. Any founder with aspirations beyond that will inevitably have to expand to new markets. The VC firm Bessemer refers to this general approach as the Second Act strategy—where companies use some combination of product innovation and market expansion to grow.

The challenge is that this second act often involves "crossing the chasm" as Geoffrey Moore put it in his eponymous book. Early startups attract innovators and early adopters; they feel the hair-on-fire problem most acutely and are willing to use a less-than-perfect product to solve it. As a startup scales, they eventually need to make the leap from forgiving early adopters to the more pragmatic and discerning majority customers. The majority market includes mainstream consumers and larger, established businesses.

So how do you cross the chasm? Moore offers another analogy: bowling. Moore calls the initial market a "headpin" market. Like in bowling, your goal is to hit the headpin and then knock down the pins behind it. The pins immediately behind the headpin are the next markets you should attack. My *Frogger* analogy from Chapter 5 works here as well; choose your preferred mental model.

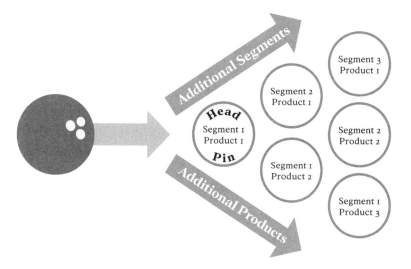

Expanding TAM Bowling Pin Analogy

Specifically, there are three ways to think about choosing the next bowling pin, lily pad, or market for scaling:

1. *Horizontal scaling:* This involves expanding into new markets or segments with the same or a similar product, platform, or service.

    C16 Bio initially focused on synthetic palm oil for niche cosmetics products but plans to expand horizontally to larger cosmetics companies and eventually into food applications. Dropbox, which started by targeting individual users for file storage and sharing, then expanded horizontally to serve small businesses and, eventually, large enterprises with essentially the same product.

2. *Vertical scaling:* This approach focuses on deepening penetration within a specific industry, market, or customer segment by adding new products and services.

   Our friends at Squire are a good example of vertical scaling. They are laser-focused on barbershops, providing products and services to them to help them run their businesses, from booking appointments to payments to ordering supplies. HubSpot is another example of vertical scaling. Its primary customer base is small and medium-size businesses (SMBs) and its initial product was a simple blogging tool. The company then built out a full-featured marketing software platform and expanded vertically to offer sales, customer service, and operations tools for its core SMB customer base.

For entrepreneurs and investors, vertical SaaS has been a goldmine. Companies like HubSpot and Squire have created tremendous value by digitizing analog industries and sleepy business processes. The combination of high gross margins (80–90 percent), recurring revenue (often net dollar retention rates exceeding 100 percent), and more efficient GTM costs (from pursuing focused target markets) has consistently generated outstanding success. Some vertical SaaS companies capture over 40 percent of their market share compared to leaders in horizontal SaaS markets that struggle to hit 20 percent, leading to better pricing power and more attractive unit economics.

3. *Geographic scaling:* This growth vector involves expanding into new geographic markets. It's a unique case worth highlighting because of the potential benefits and challenges. Geographic expansion domestically (particularly in the US) and globally is a great way to expand TAM, especially if you pick a solid macroenvironment—for example, a growing region with a stable currency and government. For example, ClassPass started with its fitness class offering only in NYC. After achieving

strong product-market fit, the company rapidly expanded to other U.S. cities and today serves customers across the country. Although there are modest differences in pricing, the offering is essentially the same in each city.

Some geographic expansions are more complicated—especially when the move is international.

Geographic Expansion: The Sprout Solutions Case Study

Founders typically look for signals before exploring international markets. Sometimes, it's obvious that your domestic market is too small or is not your ideal initial market demographically—as is the case with Israeli startups who often immediately launch in the U.S. market.

Sometimes, a startup may have a reasonably-sized domestic market to get started in, but then considers international expansion after that initial "headpin" is knocked down. Such was the case with Sprout Solutions, a company based in the Philippines that builds HR and payroll software.

Sprout was founded in 2015 by married couple Alexandria and Patrick Gentry, after an incubation period inside of a business process outsourcing firm where Patrick was a technology executive.[46] Patrick found that software tools he had built internally to help manage payroll had the potential to take off outside the company. In a few short years after spinning out, Sprout had become the largest HR SaaS platform in the country, with over 150,000 users across over 550 client companies. Sprout became profitable in 2021 and reached $6.3 million in annual recurring revenue in 2022—all while serving customers only in their domestic market.

In July 2022, the Gentrys and their team faced a pivotal decision. Having built Sprout into the largest HR SaaS platform in the Philippines, they had to decide whether to expand vertically within the Philippines with new SaaS, fintech, and even health insurance products, or to expand their TAM geographically by entering other Southeast Asian markets like Thailand, Vietnam, Malaysia, and Indonesia with their core product offering.

The decision was not an easy one. On the one hand, these neighboring countries offered tempting opportunities with similar or larger GDPs than the Philippines. Expanding into these markets could accelerate Sprout's growth and make the company more attractive to international investors, who often favor companies with global ambitions.

On the other hand, Sprout had achieved remarkable success in the Philippines and built a strong brand. The cohort behavior of their user base—which showed greater than 100 percent net dollar retention even through the COVID-19 pandemic—indicated that they had found extreme product-market fit. The startup had consistently surprised investors with its ability to scale within the Philippines, suggesting there might still be significant untapped potential in its home market. "We were getting to be the leading HR platform in the country. If we stayed local, the next step was to transition beyond HR and become the B2BOS (Business-to-Business

Operating System) for midsize businesses in the Philippines," said Alex Gentry.

As they weighed their options, the Gentrys considered several key factors:

- Market size and growth potential in the Philippines versus other Southeast Asian countries
- The competitive landscape in potential expansion markets
- The need to validate their product-market fit in new countries
- The resources required for international expansion and the potential distraction from their core business
- The expectations of current and potential investors regarding their growth strategy

Ultimately, Sprout decided to remain focused on the Philippines. This decision aligned with their original mission and allowed Sprout to build on their existing success, deepen their market penetration, and expand their services to current clients. It also positioned them well when capital availability tightened in 2023 due to rapidly rising interest rates around the world, and the company was forced to shift its focus from growth to profitability.

For Sprout, doubling down on their domestic market and expanding vertically was not the end of the story. They spent another year or two executing that strategy successfully, raising a $10.7 million Series B funding in April 2023. Then in June 2024, they made their move overseas and entered the Thailand market through an acquisition of a local cloud-based payroll and practice management software company.

TAM Expansion Sequencing

An ambitious company will eventually execute on all three expansion strategies—vertical, horizontal, and geographic—it is just a question of intentional, thoughtful sequencing and prioritization. I was in a board meeting recently where the CEO shared that they had achieved $50 million in annual revenue and serviced over 200 customers, but he was concerned that the company was "running out of TAM." He outlined a series of initiatives to develop new products, approach new customer segments, and enter new geographies. At the end of the board meeting, I asked: "All this sounds great and terrifically promising. But as we all know, we have a limited envelope of time and resources, requiring us to focus. So, which scaling vector are we going to choose first?"

Deciding on your "second act" looks a lot like choosing your initial market: You conduct research, form a hypothesis, and then run small experiments within those markets to test the fit. Look for markets similar to your current one, where your existing capabilities can add value and the investment to serve that adjacent market is modest. Assess the market size and growth potential to ensure the opportunity justifies the move. Consider the competitive landscape and your potential for differentiation. Evaluate the regulatory environment and any barriers to entry. Finally, gauge the market's readiness for your solution. Are there enough innovators and early adopters willing to work with the new product? Is the value proposition at "hair-on-fire" levels?

Then perform the same analysis within your existing market to find vertical expansion opportunities, looking at the workflows immediately before and after the one you currently solve. What are the best-in-class solutions for your customers? Where are there still massively painful problems? Think about Squire: they started with just appointment scheduling, then they moved into payments and financial management.

The more specialized an industry—such as barbershops—the more it makes sense to expand vertically, assuming the market as a whole is large enough to match your ambition.

Developing Partnerships and Ecosystems

As startups scale, strategic partnerships can be crucial in supplementing or supporting various aspects of the business, including team and talent, technology, operations, and go-to-market (GTM) strategy.

Partnerships can provide access to resources, expertise, and markets that would be challenging or time-consuming for a startup to develop independently. For example, Shippo's integration with Shopify's platform significantly expanded its reach to small ecommerce businesses. Companies that target enterprises can partner with consulting firms to help with project implementation or to be "designed in" as part of a larger solution.

Partnerships with larger companies can also support the development of scalable processes. By collaborating with established companies, startups can learn best practices and potentially adopt proven systems. A startup might partner with a larger company to leverage its supply chain or customer service infrastructure. One founder friend told me that his partnership with a Fortune 500 company almost killed the organization because of their demands, but eventually forced the company to "grow up" and be more mature and resilient.

Technology partnerships can accelerate product development and enhance offerings. These partnerships allow startups to integrate complementary solutions quickly without significant in-house development, allowing companies to "buy" versus "build" to add new features. In effect, smart partnerships—fueled by a business development team—can create an ecosystem around a company that creates an amplified, propelling force forward.

As a startup integrates additional features to enhance the user experience and broaden the product footprint, it transforms its offering from an initial feature set to a full-fledged product. This stage is then often followed by transitioning from a product to a platform:

Feature ⟶ Product ⟶ Platform

Startups can achieve this expansion successfully by opening up APIs—similar to what Shopify did in attracting Shippo to their platform—and fostering a community that can innovate on top of their core offering. By fostering an ecosystem where users, developers, and partners actively participate and contribute, startups can build a true platform that leverages community-driven innovation and exhibits strong network effects. Platforms and communities are harder to dislodge than products, thus building out a platform creates a strong barrier to entry for competition.[47]

Scaling go-to-market is one of the biggest challenges for startup founders. AI, in this context, is a double-edged sword: it gives your sales and marketing teams superpowers, but it can also lead you down the wrong path at lightning speed if you're not careful.

This leads us to our final discussion on scaling: scaling people. Choosing *who* to hire to run your AI-forward organization is the most important decision a founder will make.

Chapter 9

Scaling People: Hiring in the Age of AI

IMAGINE IF EVERY PERSON you hired was worth $10 million to your company. In other words, if you hired the right person, you'd earn an additional $10 million. If you hired the *wrong* person, the opportunity cost would be $10 million.

Would this change your hiring process? Would you think more carefully about who you hired and how you set them up for success?

This is a hypothetical question, but not for long. Revenue-per-employee metrics are skyrocketing across the tech industry as companies become leaner and more efficient with AI. Let's do a quick comparison:

Ovia Health, at the time of their acquisition in 2021, reached $20 million ARR with 100 employees. That's $200,000 in revenue per employee—a good number for a growing company in pre-gen AI days. Today, Flybridge has a portfolio company that just reached $40 million ARR, in less than a year, with just twenty employees. That's $2 million in revenue per employee—*a 10x increase*.

And we're still in the early innings of this super-scaling revolution. One founder friend, who recently raised his seed round from Sequoia, promised the firm that he would never grow to over 100 employees. He's committed to reaching a billion dollars in revenue with no more than 99 employees by incorporating AI into every part of the company—$10 million in revenue per employee.

The question for founders today is this: If each new hire is worth 10–50x more than they were previously, how do you ensure you hire the very best? What does a $10 million employee look like?

The 10x Joiner

There used to be a rule at Flybridge that any AI startup we invested in needed to have a co-founder with a PhD in machine learning. In 2012, I wrote a cheeky blog post with a nod to the classic movie *The Graduate* titled, "Hey Graduates: Forget Plastics—It's All About Machine Learning," encouraging young people to study this important field in the era of Big Data.[48]

We've scrapped that requirement today because a PhD is no longer needed to build an AI-forward company. The technical playing field is being leveled for all but the most bleeding-edge companies. There are trillions of dollars up for grabs for fast-moving startups that can "AI-ify" industries with today's existing technology, no doctorates needed. This generalizing of startup teams will only continue as AI technology matures.

It has never been more valuable to be a flexible, fast-moving generalist who can quickly learn new tools, adapt to changing circumstances, and connect dots across disciplines. The folks that can do all this *and* are AI-native are what I'll call 10x Joiners.[49]

Like the 10x Founder, the 10x Joiner can get ten times more done than a traditional startup employee, across a wider range of job functions:

- The 10x salesperson can do business development, nurture deals, and systemize the sales process better than a team of ten typical salespeople.

- The 10x marketer can design a go-to-market strategy while also deftly managing social, SEO, and PR with a team of AI agents.

- The 10x chief of staff can use specialized legal AI to write contracts and handle HR.

- The 10x (or should we say 100x) developer can turn a six-month project into a six-day project.

We already know that AI agents will end up performing 80 percent or more of our day-to-day work. A key skill in tomorrow's startups will be knowing how to orchestrate hundreds, even thousands of agents to work in collaboration. This skill applies well beyond engineering. Tomorrow's best marketers, financial analysts, salespeople, and customer support reps will all know how to coordinate with AI agents to get more done.

So as a founder, how do you screen for a 10x Joiner in the hiring process? Ask candidates to demonstrate how they are using AI copilots, and increasingly agents, in their work and lives today. Has using AI become a habit? Is it their first step in every new project? Has it become an extension of their working selves?

I recently had an interesting discussion with one of my students on the cusp of graduation. We imagined a job candidate who not just displayed high proficiency with AI, but had built a team of AI agents that she brought to the job. She would be an individual contributor *and* a manager of AI agents that expanded her capabilities, giving her a better chance of being a $10 million employee. I believe this could be the model for 10x Joiners moving forward.

Wharton Professor Ethan Mollick, author of *Co-Intelligence: Living and Working with AI*, made a similar point in a recent newsletter:

> "Given that AIs perform more like people than software (even though they are software), they are often best managed as additional team members, rather than external IT solutions imposed by management."[50]

10x Joiners don't just use AI—they build and manage hyper-efficient AI teams. Which begs the question, why hire humans at all?

Simply put, startups still need human judgment. I love the book, *Zen and the Art of Motorcycle Maintenance* by Robert Pirsig. It's a novel about the philosophical exploration into the meaning of quality versus commercialism. One of my favorite lines is about the limitations of the scientific method:

> "You need some ideas, some hypotheses. Traditional scientific method, unfortunately, has never quite gotten around to say exactly where to pick up more of these hypotheses . . . Creativity, originality, inventiveness, intuition, imagination . . . are completely outside its domain."[51]

AI is dramatically expanding our capacity to run experiments and build products, but startups still face time and resource limitations, and thus need to uncover the right answers quickly. We still need humans with creative instincts and good judgment to generate good hypotheses in the first place. Of course, these traits must be balanced with the basic competencies required for the role. But in a world where AI can increasingly handle specialized tasks, bet on the humans who can do what AI cannot: innovate, create, adapt, and thrive in ambiguity.

Hire for Jobs to Be Done, not Titles

When it comes to hiring, most founders think about *roles to be filled*—the business development representative (BDR), the VP of Engineering, the head recruiter, the social media marketer. Yet the scope of specific roles and job titles has always been ambiguous. Now with the rapid acceleration of AI, roles and titles are changing faster than ever. If you can hire one person to fill ten different roles, what would that person's title be?

The answer: It doesn't matter. I've long argued that startup founders should eliminate titles altogether to stay scrappy and agile.[52] As roles and responsibilities rapidly evolve in a startup's fluid environment, titles can become a hindrance, creating an unnecessary focus on hierarchy and scope rather than collaborative problem-solving. Establishing role clarity through *Jobs To Be Done* rather than titles ensures that the startup can adapt to changing demands and leverage AI tools effectively.

Jobs To Be Done (JTBD) is a way of thinking about different roles and functions of a business. It was developed by the late Harvard professor and my colleague, Clayton Christensen. In the JTBD model, every product, service, and role is viewed through the lens of the end user. What job are they hiring your company to do? People don't just hire a roofing company—they want their leaky roof fixed. They don't just hire a marketing agency—they want to attract more customers. So what if robots could do those jobs? If a small business could get a custom website built by AI in hours rather than weeks, does it matter who, or what, created it? Topline Pro is proving that it does not. They have focused on the *job to be done*—marketing and customer acquisition—and developed a revolutionary way to do the job, to the enormous benefit of the end user.

JTBD applies to roles within a company as well. What jobs need to be done within your startup? You need to build the product, acquire customers, satisfy customers, and make money. If those four jobs are done (which, not coincidentally, sound a lot like the four sides of the diamond in the diamond-square framework), you will build a successful business. The titles within your company don't matter.

Christensen developed the JTBD framework decades ago, but it is more relevant than ever. The jobs to be done have not changed much, but *how* we complete those jobs is evolving every day thanks to AI. Founders should be rigid on *what* jobs need to be done and flexible on *how* they are done. Do you really need a team of business development reps to send cold emails, or can you use AI to personalize outreach at scale? Does a custom software project really require six months and six engineers to complete, or six days, one engineer, and 3,000 AI agents? Everything about *how* a job is completed is changing, and will change again by the time you read this.

The *how* will change depending on the stage of the startup as well. To build the right product, a nascent-PMF startup needs to stay in constant communication with its end users, while a strong-PMF startup is more focused on scaling its product operations. The job of acquiring customers is handled very differently for early-stage versus growth-stage startups. Flexibility, again, is key, and the humans on your team will need to adapt quickly as their jobs change.

Using AI in the Hiring Process

AI can streamline recruitment by automating résumé screening, scheduling interviews, and conducting initial candidate assessments. This can significantly reduce time-to-hire and allow human recruiters to focus on high-value interactions with candidates.

One of Flybridge's portfolio companies, BrightHire, offers an AI copilot for hiring and interview planning. They improve the quality of interviews and hiring decisions by using AI to build hiring plans, improve interview consistency, save time with automated interview notes, and return actionable talent insights. Additionally, at the end of each interview, the software produces an interview quality score customized to the company's standards and the job description. By giving hiring managers visibility across interviews, they enable companies to create a standardized evaluation process, reducing bias and improving the reliability of assessments.

General AI tools can help with hiring as well. A founder friend posted a new job and received over five hundred résumés. He quickly fed ChatGPT the job description, a summary of the company culture, and the company website. Then he asked ChatGPT to assist him with filtering the résumés and helping him select the twenty candidates he should interview.

As in the case of many other areas, AI can be a great tool to assist with hiring, but founders must be cautious about overrelying on it. The human element remains crucial, especially in assessing cultural fit and soft skills. There are two excellent books on startup hiring that I recommend to all founders. The first is *Who* by consultants Geoff Smart and Randy Street, which focuses on creating a system for hiring using scorecards and conducting rigorous interviews. The second is from the former COO of Stripe, Claire Hughes Johnson, called *Scaling People: Tactics for Management and Company Building*, which provides a playbook for communication and team management.

Growing Pains: Blitzscaling Versus Blitzfailing

Reid Hoffman, co-founder of LinkedIn, introduced the concept of "blitzscaling" as a strategy for rapidly scaling a company to achieve market dominance.[53] Hoffman argues that in specific markets, especially those with strong network effects, the first company to achieve significant scale can create nearly insurmountable barriers to entry for competitors.

However, blitzscaling carries significant risks. When executed poorly or in inappropriate contexts, it can lead to "blitzfailing"—rapid, catastrophic failure due to premature scaling. Hoffman emphasizes that blitzscaling should only be attempted when the market opportunity justifies the risks, and the company has achieved product-market fit. The decision to blitzscale often involves a delicate balance between the cost of capital and the potential benefits of network effects. In markets with strong network effects, such as social media platforms or marketplaces, the benefits of rapid scaling can outweigh the high costs of aggressive growth. In markets that do not have strong network effects, or where the competition is less fierce, the opposite might be true.

TikTok is an example of a startup that successfully blitzscaled. Launched by its Chinese parent company ByteDance in 2016, the social media platform invested early in influencer marketing and product development as well as (you guessed it!) AI-based algorithms to provide personalized content feeds matched to user preferences and interests. By 2021, the company had reached an astonishing one billion monthly active users. Many of my former students work at the company, and although the outside results may look magical, on the inside there is plenty of scrambling and chaos. But if TikTok had scaled more judiciously and carefully, it might never have achieved this degree of outsized success.

WeWork, the co-working platform, is a cautionary tale of blitzfailing. The company and its investors overestimated the strength of the network effects. They also underestimated the true capital costs of scaling WeWork's business model, which depended on physical spaces and capital-intensive build outs. In 2018, WeWork filed for an IPO and for the first time, the public filing revealed the details of the company's business model and profit formula. It wasn't pretty. In 2019, the company laid off thousands of employees and returned to the drawing board to reconfigure its business model for profitability. The company filed for bankruptcy in late 2023, despite being valued at its peak at $47 billion and raising over $20 billion in capital.

Balancing Speed and Ethics

A startup that operates as an effective Experimentation Machine is well-positioned for successful scaling. The experimentation method is valuable because it gives founders a deep understanding of its customers, market dynamics, and operational requirements. When it comes time to scale, they can do so more confidently and efficiently, having already validated their core assumptions and identified potential pitfalls. AI acts as an accelerator for scale, giving startups enormous leverage to grow quickly while spending less.

While rapid scaling can lead to tremendous success, it can also amplify ethical issues if not managed carefully. In the final chapter, we'll consider the legal and ethical decisions that founders must grapple with as their companies grow and their impact on society expands. Our goal is to build companies that achieve scale and do so in a way that creates positive value for all stakeholders.

Chapter 10

Responsible AI: Embracing Ethics, Diversity, and Social Equity

I HAVE NOW GONE ON about the enormous potential of AI for nine chapters without a substantive discussion on the risks and drawbacks. That will be the focus of this final chapter. The potential downsides of AI are very real and could exacerbate some of the biggest challenges of our time. How do we embrace AI while remaining good stewards and creating a net-positive impact on the world?

Founders today must grapple with this question while also staying vigilant to avoid all the traditional pitfalls of rapid growth. Speed has always been a doubled-edged sword; add in acceleration from AI and the blade is sharper than ever.

More AI, More Problems

Generative AI represents the largest advancement in computer technology since the Internet, but it also amplifies many of the most serious issues that have come with the Internet Age. There are three main challenges to address: AI's energy consumption and environmental impact; data privacy and copyright infringement; and algorithmic bias exacerbating inequality. I generally believe there are workable solutions to each of these problems. Let's go through each one.

AI Energy Consumption: Climate Killer or Clean Energy Catalyst?

Energy usage in the U.S. has remained surprisingly flat for the last several decades despite a rise in GDP and a slew of energy-consuming products. Technologies like LED lightbulbs, hybrid vehicles, and power-efficient microchips have delivered enormous efficiencies to keep our energy output stable. Now AI is poised to disrupt this status quo.

Generative AI computing requires enormous amounts of energy; for example, the average ChatGPT query uses about ten times the energy of a Google search. To keep up with the insatiable demand for AI computing, OpenAI and other leading technology companies are rapidly building new data centers, driving the first major increase in energy usage since the turn of the century. Data centers are quickly becoming a larger piece of the energy pie. In 2022, U.S. data centers used about 3 percent of the country's power; Goldman Sachs expects that number to grow to 8 percent by 2030—a 160 percent increase.[54] This growth comes at a cost to U.S. consumers as well as the environment. U.S. utility companies will need to invest at least $50 billion in infrastructure upgrades to handle new energy generation needed for data centers. Goldman Sachs also estimates that increased carbon dioxide emissions will generate a "social cost" of $125–140 billion in the U.S.

Many tech companies are working hard to maintain energy efficiency, but they are likely to miss their short-term emissions forecasts. This rise in energy consumption has raised the alarm for environmentalists and climate change experts. They fear AI energy consumption will accelerate the rise in global temperatures and hasten the arrival of irreparable damage to our planet. The situation seems dire, but there is a silver lining.

Here's the potentially good news: AI computing is driving demand to develop clean and plentiful renewable energy. The dream since the dawn of the nuclear age has been "energy too cheap to meter." This dream was snuffed out when Americans stopped building nuclear power plants in the 1970s, but AI may be the catalyst to finally make this promised land a reality. Google, Amazon, Meta, and OpenAI are all planning to open nuclear-powered data centers for AI computing. OpenAI has even pitched the White House on opening multiple five-gigawatt data centers around the country. Five gigawatts is enough energy to power three million homes and requires the equivalent of about five nuclear plants.[55] This scale of energy generation has never been attempted; it would require an overhaul of U.S. energy policy and new investment in nuclear, renewable, and battery infrastructure. But if OpenAI can pull it off, it could spark a clean energy revolution. The United States has the second oldest power grid in the world behind Europe; AI could be the catalyst we need to invest in America's future energy security, sparking a cascade of positive effects around the world.

There is also a massive, hair-on-fire opportunity for founders (and NVIDIA, for that matter) to develop more energy-efficient AI models. Working alongside big companies, investors, and governments, startup founders brave enough to tackle the issue could change their fortunes and the world for the better.

Data Privacy and Copyright Infringement: Fair Play or Theft?

In December 2023, *The New York Times* announced a lawsuit against OpenAI and its largest shareholder, Microsoft, for using *NYT* news articles in OpenAI's training data. It is the highest profile suit in a much larger battle between AI companies and the publishers, artists, and writers who create original material. *The Times*' allegations outline the general contours of the battle:

- *The Times* alleges OpenAI used copyrighted and licensed material to train its LLMs without paying for it, amounting to theft of *The Times*'s work.

- ChatGPT occasionally memorizes whole passages, thereby committing plagiarism by recreating near-verbatim responses.

- Even when it doesn't plagiarize, ChatGPT provides such in-depth answers that it allows readers to effectively circumvent *The Times'* paywall.

As compensation for these infractions, Harvard Law Review reports, "*The Times* asks not only for monetary damages and a permanent injunction against further infringement but also for 'destruction ... of all GPT or other LLM models and training sets that incorporate *Times* works.'"[56]

OpenAI's defense is arguing the copyrightability of *NYT* articles, but it's clearly trying to avoid additional lawsuits. Microsoft has begun signing licensing agreements with major publishers to legally use their work, including a new deal with the book publisher HarperCollins.[57]

This lawsuit will not directly impact every AI-forward startup, but every founder should understand the legal implications and take measures to keep their companies safe.

First, when building your own AI training sets, for either RAG (retrieval-augmented generation) models or small language models, it's imperative you get the proper licensing for source material and protect that material from leaking into the public domain. Let me give you a simple example from my own work. In 2023, I created an AI copilot for my Launching Tech Ventures class called ChatLTV. ChatLTV was trained on my fifty-plus HBS case studies, my previous two books, class slide decks, teaching notes, transcripts from my online course, and blog posts I have written for the past twenty years. The training data included over 200 documents and 15 million words.

Because Harvard has copyrights on much of my work (and I have copyrights on my blog posts and books), it was critical that we kept the training data private. To accomplish this, we built ChatLTV using Microsoft Azure's OpenAI service rather than OpenAI's own APIs. Microsoft Azure has strong security, data privacy, and compliance standards; it guarantees that the data fed into the service is not used to retrain public models. To store the content, we use a SOC2 Type II compliant database,[58] and we pull only the relevant paragraphs from the training content to answer queries. It might have been easier to use OpenAI's direct APIs, which have gotten more secure since I first built this chatbot, but added security precautions were worth the extra time spent.

Many AI-forward startups are attempting to disrupt industries with hyper-sensitive customer data, such as financial, health, and biotech. With generative AI, these startups hope to automate manual workflows and create better, less expensive, and more personalized experiences. But the risks are serious. Exposing sensitive customer data to public LLMs could cause a massive security breach and leak data worth literally a fortune. The question for startup teams in these industries is whether they should trust public models (and build necessary safety precautions) to save time, or to build their own models on open-source software to maximize security.

Of course, the decision depends on a startup's aspirations, their technical expertise, and their customers. One Flybridge portfolio company is using AI to accelerate drug discovery for biopharma clients. They made the decision to build their product on public models and add strong security measures similar to those we used to build ChatLTV.

Data security is never a sure thing, but it's reasonable to expect that good, thoughtful product design can mitigate most security issues even when building on the major AI platforms.

Algorithmic Bias: Is the Camera Racist?

Finally, we need to address a very human problem that will be exasperated by AI if we're not careful: racism, prejudice, and discrimination. AI is trained on the data we feed it. Naturally, if that training material is biased, the results will be too. There are already many examples of racist and sexist AI outputs, but these issues can be corrected if we have the right people in the room to work on these tools. One example of "right person, right room" is Bertrand Saint-Preux, formerly a product manager at Snap.

Growing up in South Florida as the child of Haitian immigrants, Saint-Preux had always been passionate about using technology to make a positive impact. He joined social media company Snap (maker of the Snapchat app) in 2019, shortly after graduating from Boston-based Wentworth Institute of Technology. I met Saint-Preux after he completed a fellowship program with Hack.Diversity, a nonprofit I co-founded to build a more equitable startup ecosystem.

After joining Snap, Saint-Preux was assigned to a project to improve the camera output quality of Snapchat's Android app. He quickly noticed a glaring issue: the algorithms for the camera software were primarily tested on and optimized for lighter skin tones.

Historically, camera technology has misrepresented people with darker skin tones due to a legacy of discrimination embedded in its development. Since the 1950s, and for many decades after, photo labs used the "Shirley Card"—named after a white female model and Kodak employee—to calibrate skin tones. That bias persisted into the digital age, affecting everything from facial recognition to image quality. When a camera's software finds darker skin, its AI-filtering algorithm overcompensates and causes either over- or under-exposed images.

Knowing this history, Saint-Preux advocated for a more comprehensive approach that tested the camera algorithms across a wider range of skin tones. "The algorithms aren't successful because they don't work for everyone," Saint-Preux said while presenting his findings to colleagues. "Not everybody that uses Snapchat looks like the folks in this room."

"The camera is racist," Saint-Preux declared in his presentation. Of course, technology can't hold racist feelings, but it can be designed to act in racist ways, even inadvertently. "It's the systems that make technology that sometimes advance racial stereotypes," he shared.

The presentation did its job. Before long, word reached the desk of Snap's CEO, who approved the project to fix the flawed camera app.

Saint-Preux would go on to lead the development of Snap's Inclusive Camera. The project involved collaborations between developers and photographers who were experts at lighting and capturing photos of people with a wide range of skin tones. The team's efforts resulted in a camera technology that improved image quality for users of all skin tones.

When Saint-Preux told me his story, I enthusiastically gushed, "Bertrand, this is amazing! Simply because of your role in the company, being in the room where it happens, and your courage to speak out, you've impacted millions of users."

"No, Jeff," he corrected gently. "Billions."

Like all technology, AI imitates its human creators. It will never be perfect, but it *can* be better. Bertrand Saint-Preux is an example of someone taking the initiative to improve AI technology for the benefit of *everyone*.

The case of Snap's Inclusive Camera is an important reminder that when AI is trained on biased data, it can amplify harmful racial, sexist, and prejudiced stereotypes. But when trained and designed thoughtfully, AI can be a force for good. For example, well-trained AI can help minimize human blind spots. By automating processes that might otherwise be influenced by unconscious biases, AI can help create a more level playing field, especially in areas like hiring. It can also help us create better customer personas to reach a more diverse audience. "You need to be able to build fast and fail fast, so starting with an ideal customer profile makes sense," Saint-Preux shared with me. "Oftentimes, though, when [founders] are starting with a customer profile, it is someone that looks like them."

AI has the incredible power of helping us empathize with people from different backgrounds and walks of life. It can help us expand our reach to serve *all* people with the specific problem we are trying to solve, not just the sliver of audience we know personally. AI could help us to finally level the playing field, which in turn would help us create better products and a better world.

Beyond AI: The Dangers of Rapid Growth

AI's greatest benefit to startups is speed. It's also the biggest hazard. What do you do when a startup grows rapidly in a harmful direction? What happens when good intentions go awry?

The dangers of blitzfailing are greater than ever. Like a pro golfer who stops their club mid-swing to reassess, founders need the discipline to step back and say, "Hold on, let's make sure we've considered all the potential consequences." If you don't, you could end up famous for all the wrong reasons.

Juul's Blitzfailure: Good Intentions Go Up in Smoke

The meteoric rise and fall of e-cigarette company Juul serves as a cautionary tale of prioritizing rapid growth over ethical considerations. Founded by two Stanford students with the stated goal of making cigarettes obsolete, Juul launched in 2015 with a flashy marketing campaign called "Vaporized" that featured young, attractive models and colorful imagery.[59]

The product took off like a rocket, and the company fed its growth with aggressive marketing and product distribution.

This youth-oriented marketing strategy quickly drew criticism. As one public health advocate warned, "We're seeing more and more irresponsible marketing of unregulated products such as e-cigarettes." Some Juul executives began to worry. Investor Alexander Asseily cautioned in an email, "We will continue to have plenty of agitation if we don't come to terms with the fact that these substances are almost irretrievably connected to the shittiest companies and practices in the history of business."

Juul's sleek design and flavored nicotine pods proved irresistible to teenagers, despite attempts to restrict their marketing. By 2019, over 27 percent of high school students reported that they were vaping.[60] Juul found itself facing hundreds of lawsuits alleging it had deliberately targeted youth. The company paid nearly $440 million to settle a two-year investigation by thirty-three states into its marketing practices.[61]

The company is still in business, but the reputational debt that the brand and its leaders and employees have accrued still haunts them. A handful of my former HBS students joined Juul and unfortunately contributed to its rise. Knowing them, I believe their intentions were good. But I also know that when you get on a rocket ship, it is hard to control it. As one of them confided to me, "In retrospect, I wish we had slowed down and thought through all the implications of what we were doing. When you're growing exponentially, and it takes time to get the data and understand the unintended consequences, the impact is compounded and can get out of control."

When "Faking It" Becomes Fraud

Most questions of startup ethics are not so black-and-white. One example is from Khatabook, the Indian financial platform provider we covered in Chapter 6. Early on, as part of the startup's GTM plan, founder Ravish Naresh ran ads on Google and Facebook that featured an image of "the Jeff Bezos of India," billionaire businessman Mukesh Ambani. The problem was that Naresh never got permission to use Ambani's likeness. When I raise this example in my HBS classroom, many students conclude that it was a harmless act. "Ambani can afford to be the subject of a few funny memes," pointed out one student from India.

Another example is the story of Home Depot's founders stocking their shelves with empty boxes because they couldn't afford to buy enough merchandise. Was this a savvy growth hack or unethical business practice? Many of our startup heroes—from Bill Gates to Mark Zuckerberg—have admitted to tactics that are in the "gray zone" of ethical behavior. But how far is too far? There's a fine line between "faking it" and fraud, and unfortunately startups are susceptible to crossing it.

Fraud occurs when a company deceives customers or stakeholders for its own financial gain. This sounds scarily similar to the "fake it 'til you make it" advice that many founders receive and follow. It's no wonder that startups are so susceptible to committing fraud. Prominent examples from the last decade include Sam Bankman-Fried of FTX, and Elizabeth Holmes and Ramesh Balwani of the failed startup Theranos. All were indicted, found guilty of fraud, and sent to prison.[62]

The fraud triangle, developed by criminologist Don Cressey in the 1970s, identifies the three driving forces that enable fraud. Startups experience each of these forces in spades and will sound familiar to any founder reading this:

The Fraud Triangle by Donald R. Cressey

The Fraud Triangle

- **Pressure:** Startups have severely limited resources and time to make their ideas come to life. There is a huge amount of pressure to "make a dent in the universe" and a culture of striving to be an outlier. This pressure encourages more risk-taking and, because of the many heroic founder success stories, the community celebrates it.

- ***Rationalization:*** Startupland encourages founders to "move fast and break things" and "fake it 'til you make it" as strategies for the underdog to unseat large incumbents in any market. When this advice comes from the most successful CEOs in tech, it's easy to justify unethical actions.

- ***Opportunity:*** There is tremendous opportunity for fraud due to immature companies with weak governance and nascent control systems. There is also a "founder-friendly" culture of trust and, as is the case with many private companies, a general lack of transparency that allows bad actors to operate with impunity.

I believe that most entrepreneurs start out with good intentions, but under pressure to succeed, some cut corners. If a founder is not careful, they can take their "growth at all costs" mantra to a dangerous and damaging extreme. Sometimes they stoop to the level of fraud, which only gets worse as they try to dig their way out before getting caught.

Fraud is exceedingly rare even in the high-pressure world of startups, but the important thing for founders to remember is that *anyone* can fall into the trap of committing fraud. As the old saying goes, "The road to hell is paved with good intentions." Your actions determine your fate, so be on guard.

Building a Better Startupland: Equity and Opportunity

Startupland has many challenges beyond the scope of AI, and it will be up to the next generation of founders to solve them. Although much progress has been made, the reality is that the startup playing field remains uneven, especially for women and other non-white-male groups. The disparities in funding and outcomes for these groups are stark, persistent, and systemic—and they can often act as unfair barriers and filters that decide who gets to enter and succeed in Startupland.

In the U.S., VC-backed startups with women founders account for only 2–3 percent of all funded ventures, while only 15–20 percent of startups have a woman on the founding team. Black entrepreneurs make up a mere 2–3 percent of the VC-backed startup ecosystem (in contrast to representing 14–15 percent of the U.S. population), and the figures for Latino founders are similar despite representing 18–20 percent of the U.S. population. What about the checkwriters themselves? No surprise, there is a correlation. Women represent only 10–12 percent of senior VCs, while Black VCs make up 1–2 percent and Latino VCs another 2–3 percent. (Sources at the end of the chapter.)

Venture capital is the lifeblood of high-growth startups. In his book *Behind the Startup: How Venture Capital Shapes Work, Innovation, and Inequality*,[63] UPenn sociology Professor (and former tech startup joiner) Benjamin Shestakofsky argues that the VC system is designed to funnel rewards to the elites, often sidelining those who do not fit the traditional mold of a startup founder. His research underscores a root cause of demographic wealth gaps: the ownership of assets, which remains concentrated among a select few, compounded by VC funding disparities.

Underrepresented founders often start with less capital and face more significant challenges in raising additional funds. Studies have shown that Black-owned startups, like Squire, are 50 percent less likely to secure small business loans compared to their white counterparts, and when they do, the amounts are typically lower.[64] This lack of access to capital hinders their ability to scale, innovate, and compete on an equal footing with other startups.

Moreover, the entrepreneurial process itself—finding co-founders, hiring talent, securing legal and financial advice—often operates within networks that exclude—or marginalize—underrepresented founders. These networks are crucial for accessing the resources needed to grow a startup, yet they are often closed off to those who don't already have the "right" connections or belong to the "right" networks. The result is a self-perpetuating cycle where underrepresented groups remain on the periphery of the startup ecosystem.

The Pipeline Myth

So how do we fix this? One way is to cast a wider net. Some argue that underrepresentation in business and tech is due to a "pipeline problem"—a shortage of Black, brown, and female talent to fund or hire. However, my experiences with Hack.Diversity and related initiatives have shown that this theory isn't true if you're intentional about finding diverse talent.

A case in point is XFactor Ventures, a fund co-founded by my Flybridge partners, Anna Palmer and Chip Hazard, to focus solely on backing female founders. When it started, skeptics raised concerns about whether there were enough qualified female founders to invest in. After the first year, Chip noticed that his calendar had dramatically shifted in the gender profile of the founders he was meeting. Before XFactor, the founders he met with were roughly 80 percent male and 20 percent female. After launching the fund, this ratio moved closer to 50/50. It was as if he had discovered a "hidden" pipeline. By building a fund specifically for female founders and recruiting a network of over twenty female CEOs as the investment team, Anna and Chip created a magnet for female entrepreneurs. This force generated an enormous amount of deal flow from talented women who might have otherwise been overlooked.

From our experience, the pipeline isn't low or leaky—it's often just untapped. When you build intentional structures and networks to reach underrepresented groups, you'll likely find that there is sufficiently qualified talent waiting for the opportunity to contribute.

Opening the Door to "The Room Where It Happens"

Warren Buffett often talks about "the ovarian lottery" to describe the significant role that chance plays in determining the circumstances of one's life. In Startupland, the conditions that shape your path to success—whether personal or entrepreneurial—are starkly different if you were born to two

Harvard-educated PhDs living in Boston (as I was) compared to being born on a small island in Indonesia (as was one of my most successful unicorn founders).

That's why it is also important to recognize the business opportunity in addressing social inequities in business. When Bertrand Saint-Preux joined Snap, he brought with him a lived experience as a Black man and a knowledge of how some groups in tech are overlooked and underestimated. Thanks to his experiences with programs like Hack.Diversity, where he now serves as our VP of Programs, Saint-Preux got an opportunity to be "in the room where it happens." Because he was at the table where decisions were being made, he was able to offer a perspective that wasn't previously considered or appreciated. As a result, Snap benefited because it now offers a better product that meets the needs of a wider user base.

Every Startupland stakeholder's goal should be to create an ecosystem that works for everyone, with an understanding that our fates are often intertwined. As Jennifer L. Williams, co-founder and CEO of Diversd (a counter-bias human resources platform) aptly put it, at the intersection of diversity, equity, and inclusion is *belonging*. "Ultimately, you want an organization where talent from anywhere can come and succeed for themselves and for the business."

The same is true for Startupland as a whole: we want to build ecosystems where entrepreneurs from anywhere can succeed.

Guiding Principles for Embedding Ethics into Your Startup

Now, more than ever, founders need to be intentional about operating under ethical principles. I like the framework for responsible scaling outlined by General Catalyst's Hemant Taneja:

- Consider the systemic societal change we aspire to create
- Sustain the virtue of our product
- Hold ourselves accountable at scale

Taneja co-founded a nonprofit called Responsible Innovation Labs that published a Responsible AI protocol,[65] a practical playbook to help founders implement a set of realistic and relevant standards.

As we approach the end of this chapter and the book, I will share a few lessons about building startups responsibly in the age of AI that I have learned over the years:

1. *Embed ethics from day one.* Make ethical considerations a core part of your startup's DNA from the beginning. Include ethics in your business plan, mission statement, and founding principles. Clearly identify the lines you *will not* cross.

2. ***Develop a clear Responsible AI framework.*** Create a structured approach to AI development that addresses key areas like fairness, accountability, transparency, and privacy. Use this framework to guide all AI-related decisions. Remember, when safety is the price of speed, don't pay it.

3. ***Balance innovation with responsibility.*** While pushing technological boundaries, always consider the potential impacts on individuals and society. Have the courage to slow down to consider the unintended consequences of your actions.

4. ***Prioritize diverse perspectives.*** Build a team with diverse backgrounds and viewpoints to help identify and address potential ethical blind spots in your AI systems and business practices. Additionally, actively seek out a demographically diverse customer base; there are always numerous points of view, even in the most niche of initial markets.

5. ***Implement ongoing ethical assessments.*** Regularly evaluate your AI systems and business practices based on your ethical standards. Be prepared to adjust your approach as new challenges emerge.

6. *Foster a culture of open dialogue.* Encourage team members to voice ethical concerns without fear. Create channels for discussing and resolving ethical dilemmas. Implement a company policy that *anyone* can bring up ethical concerns and they will not be punished or ostracized for it.

7. *Translate principles into actionable guidelines.* Move beyond abstract ethical principles by developing specific, practical guidelines that team members can apply in their daily work. This includes hiring practices, AI training, financial transparency, and data privacy protocols.

8. *Engage with external stakeholders.* Collaborate with academia, industry peers, and regulatory bodies to stay informed about emerging ethical standards and best practices in AI. Invite them into your company as a source of accountability.

9. *Prioritize transparency with users.* Be clear and honest with users about how your AI systems work, their limitations, and how user data is used. Build trust through openness.

10. *Plan for responsible scaling.* As your startup grows, ensure that your ethical practices

scale with it. Develop tools and processes that allow new team members and departments to easily adopt and implement your responsible AI practices. Don't be afraid to slow down.

Ethics, diversity, and social equity should be integral to a founder's vision and values. As much as speed is valuable in the startup context, there's also wisdom in the saying, "If you want to go fast, go alone; if you want to go far, go together."

By prioritizing and celebrating ethics within our ecosystem, we not only foster a more inclusive and just environment but also ensure the long-term success and sustainability of our ventures. The good news is that, when done correctly, we can prioritize both—that's the secret sauce of sustainable growth and enduring brands.

Reference/Reading List

Jeffrey Bussgang, "Are VCs racist? Explaining the capital gap," LinkedIn, January 3, 2022, https://www.linkedin.com/pulse/vcs-racistexplaining-capital-gap-jeffrey-bussgang/.

Ewens, M. (2022). Gender and race in entrepreneurial finance. https://doi.org/10.3386/w30444.

Responsible Innovation Labs. (n.d.). *Responsible AI*. https://www.rilabs.org/responsible-ai.

Gorman, James. "The Growing Market Investors Are Missing," 2018. https://www.morganstanley.com/content/dam/msdotcom/mcil/growing-market-investors-are-missing.pdf.

Davis, Dominic-Madori. "Funding for Female Founders Remained Consistent in 2023." *TechCrunch*, February 21, 2024. https://techcrunch.com/2024/01/10/funding-for-female-founders-remained-consistent-in-2023/.

Metinko, Chris, and Teare, Gené. "Drop in Venture Funding to Black-Founded Startups Greatly Outpaces Market Decline." Crunchbase News, November 7, 2024. https://news.crunchbase.com/diversity/venture-funding-black-founded-startups-2023-data/.

Kunthara, Sophia. "Black Women Still Receive Just a Tiny Fraction of VC Funding despite 5-Year High." Crunchbase News, August 23, 2021. https://news.crunchbase.com/diversity/something-ventured-black-women-founders/.

Abouzahr, Katie, Matt Krentz, John Harthorne, and Frances Brooks Taplett. "Why Women-Owned Startups Are a Better Bet." BCG Global, February 3, 2023. https://www.bcg.com/publications/2018/why-women-owned-startups-are-better-bet.

Morgan Stanley. "Mixed Signals on Equity in VC Funding" https://www.morganstanley.com/ideas/vc-funding-diverse-startups-2022-survey.

Solal, Isabelle, and Kaisa Snellman. "For Female Founders, Fundraising Only from Female VCs Comes at a Cost," February 1, 2023. https://hbr.org/2023/02/for-female-founders-only-fundraising-from-female-vcs-comes-at-a-cost.

Conclusion

The Experimentation Mindset

THROUGHOUT THIS BOOK, I have stressed the need for speed in startup experimentation and the role of AI as an experiment catalyst—a turbo boost for finding product-market fit. We are in a unique moment in history where the tools for founders have never been more powerful. This is truly the best time ever to be an entrepreneur.

This book has given you a framework for building startups in the age of AI, but it is not an exact blueprint. To execute successfully—and yes, swiftly—you need to have an experimentation mindset. It's a mindset that optimizes for incisive, nuanced insights and strategic thinking—not just the execution of a good process or tactics.

The best scientists and mathematicians possess an experimentation mindset. They challenge, observe, question, test, and iterate; and then they do it again and again. Why? Because they are obsessive, habitual learners. They are themselves experimentation machines. They understand that learning compounds and each new insight builds on the last. Good judgment leads to good experiments, which leads to good outcomes... which in turn leads to better judgement, experiments, and outcomes.

When early-stage founders focus on learning over vanity metrics and even profit, they set themselves up for finding genuine (not premature) product-market fit in the near term while establishing a culture of experimentation and learning for the long term.

Here are a few parting thoughts on developing an experimentation mindset as you begin your journey:

Strategy Is Old-Fashioned Human

Brian Elliott and Sid Pardeshi, the co-founders of Blitzy.AI, are building a software development platform that can shrink a project from six months to six days. A question they get often is, "If AI can do so much, why do we need humans in the mix at all?"

Their answer: *Strategy* will always be old-fashioned human.

As artificial intelligence takes over more of the *what* and even the *how*, humans must become experts in the *why*. Strategy is deciding on:

- Your company's aspirations
- The problem you'll solve
- The customer you'll serve
- The product you'll create first

- The go-to-market plan
- The business model and profit formula

Most of all, strategy is deciding which tests to run, and when, to develop the best course of action for each of these steps. The most valuable skill for founders today is not a technical capacity; it's sound judgement, clear communication, strategic thinking, and an experimentation mindset.

Taste or discernment is another innately human trait that AI will struggle to replicate. Christopher O'Donnell, former head of product at HubSpot and founder of AI CRM startup, Day.AI, colorfully put it like this to me: "AI is very helpful in that it can design a thousand pairs of sneakers, but it can't wear the sneakers for you, and it can't tell you which sneakers are cool."

AI can help us expand and scale good ideas, but it's up to us, the humans in the room, to determine which ideas are worth pursuing.

Embrace Your Inner Edison

Developing an experimentation mindset requires approaching your venture like a scientist. Every new idea or strategy starts as a testable hypothesis. Whether it's a product feature, a marketing campaign, or a revenue stream, it begins with an educated guess: "If we do X, we expect Y to happen." This clarity helps set the stage for experimentation.

Unfortunately, I see many founders make vague and untestable guesses—no wonder they end up with inaccurate insights that put them on the path to startup failure. Starting with a testable hypothesis, informed by research, encourages critical thinking and ensures that actions are data-driven, not wild guesses.

Once the hypothesis is in place, running small, manageable experiments is next. The bulk of this book is dedicated to the essential experiments all founders need to run to find product-market fit. The key is to perform experiments quickly and efficiently, which is where AI becomes a big advantage. Build an MVP and use it to collect data until you see a pattern emerge.

Startups can be like roller coaster rides, full of highs and lows. Be careful to stay laser focused on the end goal. Whatever the result of an experiment, treat it as an insight that prompts an iteration. Either repeat the experiment or formulate a new hypothesis and run a new experiment.

Don't Wait for AI to Mature

Despite the flood of generative AI updates since ChatGPT debuted in 2022, the truth is that the technology is still evolving. As Ethan Mollick notes, "Whatever AI you're using right now is the worst AI you're ever going to use." This fact makes some founders reluctant to experiment with current, sometimes buggy AI solutions.

Waiting for AI to mature is a mistake; instead, founders should jump in and get their hands dirty. Founders who start experimenting with AI now will gain valuable experience and proficiency that late adopters will find difficult to match. Start by exploring new AI tools, applying them to different use cases, and seeing what works. Over time, this AI-forward experimentation will become a deeply ingrained organizational habit, which in turn will allow the organization to adapt the next wave of AI tools faster than the competition. While getting started is hard due to inertia, overcoming this challenge is crucial to becoming adept with AI.

As I said at the beginning of the book, AI may not replace founders. But founders who learn how to use AI effectively will outperform founders who don't.

Founders who fear they've already missed the AI train should recruit a few "reverse mentors" to teach them—perhaps someone less experienced than you, but already facile with AI. Ask them to teach you the ins and outs of prompting, building custom chatbots, deploying AI agents, and more.

Or if you'd rather learn on your own, ask AI to teach you how to do tasks with AI. Try this prompt:

> *"I am doing [insert task or project] and want to use AI to automate some of the work, but I don't know where to start. Please give me some ideas on how to use AI in this capacity."*

You'll receive a list of ideas; most good, some not. Pick an idea that seems promising and then ask, "Give me a step-by-step breakdown of how to build this."

This simple prompt will open up a whole new world of possibilities for you. It will show you—as my HBS students have learned—that you can build virtually anything regardless of your level of technical knowledge. The greatest limitation is your own imagination.

Get One Percent Better Every Day

The goal isn't to become an AI expert overnight. Start small, test new tools, and build from there. Each experiment with AI is an opportunity to learn, refine your approach, and discover new use cases.

In his instant classic, *Atomic Habits*, author James Clear observes how small improvements compound over time. In the book, Clear introduces the "1 percent rule," which states that if you improve by just 1 percent daily, you'll be 37 times better by the end of the year.[66]

If you choose to work with AI every day, those small gains will add up. Maybe the outcome will be 37x. Perhaps a little less—or a lot more.

Leverage AI to Become a 10x Founder

Throughout this book, you have met incredible founders who have experimented their way to startup success. Some of these founders, who are building their startups in the dawn of the AI age, are experiencing dramatic gains in their productivity and speed. They are experimenting with AI in every aspect of their business, from ideation to customer discovery, MVPs to GTM plans, product features to people management. They refuse to let the limitations of the past determine their future.

They recognize the paradigm shift they are living through and are fully embracing what's on the other side.

Are you ready to join their ranks? Are you ready to become a 10x Founder who's not just riding the wave, but shaping it?

The future belongs to the experimenters. What are you waiting for?

Appendix A

AI Prompt Tips and Examples

THIS BOOK CONTAINS MANY prompt examples. The purpose of this appendix is to step back and provide you with some advice on how to be effective in communicating with LLMs—that is, the art of prompt engineering.

Prompt engineering refers to the skill of formulating inquiries or instructions to AI in a way that maximizes the effectiveness, relevance, and usefulness of its responses. It requires an understanding of the capabilities of LLMs as well as the nuances of crafting precise and unambiguous prompts. The better you are at communicating with AI, the more you can leverage its capabilities.

My orientation is to have longer prompts than shorter ones, with more detail rather than less. In general, the amount of text an LLM can consider at once—known as the "context window"—has gotten larger and larger. This trend allows more relevant information to be passed in the prompt, including entire documents. Even though it may feel pedantic, communicating with AI is an exercise where being pedantic is a feature, not a bug.

My other overall advice is once you read through these tips, experiment with what works best for you. Treat the AI like a team member. Or as Wharton Professor Ethan Mollick likes to put it, think of AI as a person—in particular, an intern. You have to onboard it with context and then give it coaching and feedback along the way. Again, don't be afraid to be pedantic and directive.

Good Prompt Components

Here are the five key components to good prompts:

Persona Definition

Defining the persona from which the AI should respond is crucial for eliciting relevant and contextually appropriate responses. This involves specifying not only your role but also the context and nuances that accompany it.

For example, instead of saying, "I am a VC," you might say, "I am a venture capitalist who invests in seed-stage AI start-ups with offices in Boston and NYC." This level of detail helps the AI anchor its responses in the appropriate professional and geographical context, thereby enhancing the relevance and accuracy of the output. Detailed persona definitions leverage the expanded context windows of modern LLMs, ensuring that the AI has sufficient information to generate high-quality responses.

For instance, in my experience, defining myself as, "a professor at Harvard Business School specializing in technology entrepreneurship, startups, and AI and teaching a course on pre–product-market fit startups to second-year MBAs who aspire to be startup founders and joiners" provides a richer context for the AI, leading to more insightful and tailored responses compared to a generic description.

Specific Task or Objective

Clearly articulating the specific task or objective you aim to accomplish is crucial when prompting LLMs. This involves not only stating what you want the AI to do but also providing enough context to guide its response effectively.

For instance, instead of a general request like, "draft an email," you should specify, "I am drafting an email to a potential seed investor outlining our new product's benefits and why it is a compelling investment opportunity." This level of detail helps direct the AI to the purpose of the task and the desired outcome, leading to more accurate and relevant responses. Detailed task descriptions leverage the AI's capability to process and integrate specific information, making the responses more targeted and useful.

For example, when you provide a prompt such as, "Outline a business proposal for a tech startup looking to secure seed funding, highlighting our team, market analysis, competitive advantage, and financial projections," the AI can tailor its response to the intricate needs of the task, producing a coherent and comprehensive proposal that aligns with your strategic goals.

Additionally, specifying the task or objective ensures that the AI stays focused on the desired outcome, reducing the risk of off-topic, irrelevant, or even outright inaccurate responses (sometimes referred to as "hallucinations"). This is particularly important in complex scenarios where precision and clarity are paramount. By setting clear objectives, you enable the AI to apply its capabilities in a structured manner, thus enhancing the efficiency and effectiveness of the interaction.

Step-By-Step Instructions

Providing step-by-step instructions, also known as "chain of thought" commands, is a powerful technique for enhancing the performance of LLMs. This approach involves breaking down complex problems or tasks into smaller, sequential steps, enabling the model to handle each part in a structured and logical manner. This methodology mirrors the way programmers are taught to tackle coding challenges: by decomposing a large task into manageable sub-tasks. When I was a young computer science major in college, this notion of breaking down a larger task into small tasks was a revolutionary way for me to think about problem-solving. Because AI works this way, it helps to work with AI using this same mindset.

For example, instead of asking an AI to "create a marketing strategy," you should outline the specific steps required to develop the strategy. This could include tasks such as conducting a comprehensive market analysis, identifying target demographics for your user persona, developing key messaging, and outlining a promotional plan. By delineating these steps, you help the AI navigate the complexity of the task, producing more coherent and actionable responses.

The chain of thought technique not only clarifies the process for the AI but also improves the quality of the output. When the AI is guided through a series of logical steps, it can generate responses that are more detailed and systematically structured, thereby enhancing the overall effectiveness of the interaction. This is particularly beneficial for tasks that require meticulous planning and execution, as it ensures that no critical components are overlooked.

Additionally, step-by-step instructions facilitate better comprehension and learning for users. When founders lay out the logic of their command in a methodical fashion and see how the AI reacts to that methodology, they can develop a deeper understanding of both the task and the AI's problem-solving capabilities. Sometimes, instead of building out step-by-step instructions in the same prompt, you might consider a multi-stage series of prompts. For example, I find that when asking AI for help with feedback on a document or piece of writing, feeding it paragraph by paragraph or page by page is more effective than feeding it the entire document.

Audience Definition

I also like to clearly tell the LLM who the intended recipients of the output are. Again, more detail is better. "Explain in detail the unique selling proposition of my new product idea to my target customers. My target customers are claims processors who are responsible for preventing fraud at healthcare plans in the U.S. These plans are the largest ones in the country who typically cover between one million and fifty million lives." Or "I want to communicate clearly and concisely to a group of ninety Harvard MBAs who are in their second year and keen to learn about entrepreneurship because they are each aspiring startup founders or joiners."

This level of detail helps the AI tailor its response to the specific knowledge level, interests, and concerns of the audience. Clearly defining the intended audience for the AI's output is crucial for ensuring the relevance and effectiveness of the response. This involves providing detailed information about who the audience is, their needs, and their expectations.

Detailed audience definitions enhance the AI's ability to produce outputs that are not only accurate but also engaging and relevant. For instance, when I am using the AI to assist me in communicating to my students, I specify, "my audience consists of second-year Harvard MBA students who are aspiring entrepreneurs keen to learn about entrepreneurship. After they complete my class, they go off and start or join early-stage tech companies." This detail helps the AI understand the context and tailor its language and content accordingly, ensuring that the response resonates with the audience and meets their specific needs.

Additionally, defining the audience helps in setting the tone, style, and complexity of the response. A detailed audience profile allows the AI to adjust its language, examples, and explanations to match the audience's background and expectations. This is particularly important when communicating with diverse groups, as it ensures that the content is accessible and impactful for all recipients.

Output Format

Specifying the output format is essential to guide the LLM in producing content that meets your exact needs. This involves detailing the structure, length, style, and tone of the desired output. For example, instead of asking an AI to list out your competitors, try something like: "List out for me the top competitors to my new product idea in the form of five bullet points. For each competitor, provide three pros and three cons regarding their solution in the voice of their customers."

Being explicit about the format helps the AI tailor its responses to the intended use, whether it's a list, a formal report, a casual email, or an emotional customer review. An AI founder friend of mine likes to add the word "emotional" in many of his prompts. For example, "For each of these top three competitive products, share the emotional reactions from their customer online reviews that stand out in terms of positives and negatives about the product." Invariably, I instruct LLMs how many sentences or bullets I want them to produce and the specific tone for the content (e.g., formal, informal, technical, creative).

As mentioned earlier, we are investors in TalkTastic, a startup that works across your apps to rewrite whatever you say to it (leveraging advancements in natural language processing and speech recognition) in a format that you specify—e.g., a LinkedIn post, a professional email, a *NY Times* article. My VC partner, who is a very auditory thinker and likes to pace around and talk through his thoughts, uses the app a dozen times a day to communicate more effectively through Slack and email.

In sum, here are the key elements of a good prompt:

1. Persona definition
2. Specific task or objective
3. Step-by-step instructions
4. Audience definition
5. Output format

Let's take an example all the way through, weaving together these five components.

Example: Create an Employee Onboarding Presentation

Suppose I am a founder who wants to create an onboarding presentation for new hires who join my startup. Here is a prompt I might provide an LLM to help me get started:

Create a detailed onboarding presentation for new hires at my startup. My target audience: new hires who are recent graduates with little to no prior work experience in our industry. The presentation should be structured to cover the following key sections:

1. **Company Overview:** Provide a brief history of the company, its mission, vision, and core values. Include information on the company's founding story, major milestones, and key products and services as well as the sector we operate in and the customers we serve. The new team members should understand from this section our unique company story.

2. **Team Introduction:** Introduce the key team members (founders, management team, investors, advisers), their roles, and how their work contributes to the company's goals. Include photos and a short bio for each team member. The new team members should understand from this section the org structure and key contacts.

3. **Company Culture:** Explain the company's culture, work environment, behavioral norms, and expectations. Highlight any traditions, social activities, and employee resources. Include examples of how the company supports work-life balance and employee professional development. The new team members should understand from this section our unique culture and how to integrate quickly into it.

4. **Day-to-Day Operations:** Outline the daily workflow, including an overview of the tools and technologies used, standard procedures, and any important protocols. Provide examples and practical tips to help new hires get up to speed and contribute immediately.

5. **Policies and Benefits:** Detail the company's policies regarding work hours, remote work, dress code, vacation policy, and any other relevant guidelines. Also, explain the benefits package, including health insurance, retirement plans, and any additional perks. The new team members should understand our policies and benefits and feel secure, taken care of, and motivated.

Ensure that the presentation is visually appealing and engaging, with slides that are concise, visual, but informative. Use a friendly and welcoming tone throughout, as if you were a cheerful but professional head of human relations. Include opportunities for questions and interaction. I am attaching our employee handbook (as a PDF document) and branding style guide (as a PowerPoint document).

By the way, an LLM can help you create your employee handbook as well as your visual style/brand guide. Feeding an LLM relevant documents with context is easy and an important way to ensure quality and accuracy.

Example: Investor Email

Founders spend a ton of time carefully drafting emails to investors to solicit their interest. AI is excellent at doing outbound emails if prompted correctly. Here is one example prompt a founder might utilize:

Draft an email to Jeff Bussgang of Flybridge Capital Partners. I want to approach Jeff as a potential seed investor. The email should outline the benefits of our new AI-driven healthcare analytics platform. The email should highlight the following key points:

- The innovative features of our product, including real-time data analysis and predictive modeling

- The significant market demand and potential growth opportunities within the healthcare sector

- Our competitive edge, supported by recent partnerships with leading healthcare providers

- Testimonials from early adopters and pilot program successes

- Make specific references to Flybridge's other portfolio companies that relate to what we are doing, specific references to Flybridge's publicly stated investment thesis in AI and use this information to articulate why we think we are a particularly good fit for them and their investment thesis.

- A call to action for Jeff to schedule a follow-up meeting to discuss potential investment opportunities. The tone should be professional yet engaging, emphasizing the strategic value and potential ROI for his investment.

In this example, you are not only asking for a straightforward email to be written, but also asking the AI to do some research for you ("other portfolio companies" and "stated investment thesis") as if it were acting as an agent on your behalf to assist you in the task. You might even consider attaching your pitch deck, customer outreach email, and perhaps a PDF with a few dozen other emails you've written for context and to help nail the right style and tone.

Example: Create a Social Media Marketing Plan for a New Product Launch

Suppose you need to create a comprehensive social media marketing plan for your new product launch, a new biodegradable diaper (a real startup!). Here is how you could provide step-by-step instructions to an LLM:

Develop a detailed social media marketing plan for the launch of our new biodegradable diaper. Attached is a description of the product in the form of a PDF. The plan should include the following steps:

1. *Market Research:* Conduct market research to identify key trends and target demographics interested in eco-friendly products.
   - Analyze competitor campaigns to identify successful strategies.
   - Gather data on potential customer preferences and behavior.
   - Define a user persona or ideal customer profile (ICP) that might represent an attractive "beachhead market." Be very specific and targeted in this description.

2. *Platform Selection:* Determine the most effective social media platforms for reaching our target audience.
   - Evaluate platforms such as Instagram, Facebook, Twitter, TikTok, and LinkedIn based on user demographics.
   - Define a series of experiments to test out each platform in terms of their CAC, LTV, and cohort behavior.
   - Justify the selection of each platform with relevant data.

3. *Content Strategy:* Develop a content strategy that includes various types of posts (e.g., images, videos, articles) and key messaging points.
   - Create a content calendar outlining the schedule for posts leading up to and following the product launch.
   - Develop a mix of educational, promotional, and engagement-focused content.

4. *Influencer Collaboration:* Identify and reach out to influencers who align with our brand values and target audience.
   - Compile a list of potential influencers that are relevant for our target beachhead market/ICP and assess their engagement metrics.
   - Draft personalized outreach messages to propose collaboration opportunities.

5. *Engagement Plan:* Outline strategies for engaging with followers and responding to comments and messages.
   - Establish guidelines for customer service interactions on social media.
   - Plan interactive activities such as Q&A sessions, polls, and giveaways to boost engagement.

6. ***Performance Metrics:*** Define key performance indicators (KPIs) to measure the success of the social media marketing campaign.
   - Set specific, measurable goals for metrics such as engagement rate, follower growth, and conversion rate.
   - Plan regular reviews and adjustments to the strategy based on performance data.

Ensure that the social media marketing plan is detailed and actionable, with each step clearly defined and justified. The tone should be professional and strategic, reflecting the importance of the product launch and our commitment to sustainability.

Again, attaching some context would help here. And, not surprisingly, a bit of back and forth (even if pedantic) might be needed to refine the output you want.

Example: Help Me Prepare for a Pitch with Investors

Suppose you have a meeting with a prospective investor in a few days and want to practice a bit before the big session. You could try something like this:

Pretend you are a Series A investor focused on investing in health technology startups. Your past portfolio companies include [insert examples]. You have a reputation for being fair but rigorous and even tough in meetings with founders. You tend to be skeptical of wild claims from founders. Roleplay an introductory investor meeting with me. Take charge of the meeting and ask me incisive questions every step of the way. Do it step-by-step. Ask one question at a time and then wait for me to submit my answer after each of your questions. Respond to my questions with a follow-up question or two for each topic to probe my underlying assumptions and don't let me get away with surface-level answers. Let's do this for ten questions. After you ask me ten questions and I respond to ten questions, give me feedback on my answers one by one. Again, be rigorous and hold me to a high standard. Attached is my pitch deck for context.

Example: Create an Expert Prompt-Writing AI

Now that we've covered the fundamentals and shared a few examples, here is a clever shortcut to help you create better prompts, faster. It comes from Abhi Dewan at Venture5 Media. He and his team created a standard prompt format to enlist ChatGPT to create expert prompts, effectively deploying AI to prompt itself. Here is the prompt Abhi Dewan uses:

> *I would like you to act as an AI Prompting expert or advisor to me, helping me create effective prompts. The prompts will be used for various purposes*

[e.g., customer support, internal training, technical guidance, etc], which is to be specified each time a prompt is requested. You will provide me with a *short* set of questions/inputs to gather relevant info for each new prompt. After you've gathered enough info, the output should be a prompt I can use to accomplish my goal. To start, feel free to ask me a few questions to learn more to help create more targeted and useful output.

Summary

The best way to learn how to be effective at prompting is to do more prompting. And don't be afraid to correct the AI if it gets something wrong. In fact, the more you do it, the more they (and you) will appreciate it! One of my favorite things to add to the end of a prompt where I am asking for feedback is something like, "Be more critical than normal. I strive for excellence and want your response to reflect my high standards." Otherwise, the AI tends to be too nice.

Speaking of being nice, one of my HBS colleagues, Professor Mitch Weiss, likes to insert the words "please" and "thank you" in all of his prompts. When he is speaking to audiences explaining why he does it, Mitch likes to say (with a wink) that he adheres to polite dialogue with the AI, "just in case." Personally, I don't worry about AI taking over the world and being in control of us humans. But just in case, you may also want to add in the occasional "please" and "thank you".

Other Prompting Resources:

Jules S. Damji. "Best Prompt Techniques for Best LLM Responses - the Modern Scientist - Medium." *Medium*, March 14, 2024. https://medium.com/the-modern-scientist/best-prompt-techniques-for-best-llm-responses-24d2ff4f6bca.

"Elements of a Prompt – Nextra," November 18, 2024. https://www.promptingguide.ai/introduction/elements.

OpenAI. "Prompt Engineering." OpenAI Platform. Accessed December 20, 2024. https://platform.openai.com/docs/guides/prompt-engineering.

Sheila Teo. "How I Won Singapore's GPT-4 Prompt Engineering Competition." *Medium*, April 21, 2024. https://towardsdatascience.com/how-i-won-singapores-gpt-4-prompt-engineering-competition-34c195a93d41.

Appendix B

Startup Valuations

IN CHAPTER 6, I share what elements comprise a high-quality business model and how to run experiments to test and learn your way toward those models. I hint at startup valuations—how is it that some startups are worth 10x revenue and others are only worth 2x revenue—but I don't provide the details behind what specific factors and methodologies are used to value startups. That is the focus of this appendix.

Foundation: Company Valuations

To understand startup valuations, you have to begin with a foundational understanding of company valuations in general.

Companies are valued by estimating their future cash flows and discounting those cash flows back to their present value. There are many assumptions that need to be made in this calculation, which is known as the discounted cash flow (DCF) method:

- What are the future cash flows? To value a company, you need to estimate and project out these figures over a long period of time, recognizing that these cash flows are highly variable and unpredictable and rest on a set of underlying assumptions about the quality and competitive position of the company and the skill of its management team.

- What discount rate to choose? The discount rate reflects the risk associated with these future cash flows and the general macroeconomic environment involving other factors such as interest rates and inflation. A high discount rate implies you don't value the future cash flows very highly. A low discount rate implies a low-risk, low interest rate, low-inflation environment in which future cash flows are highly valued.

In Chapter 6, I shared the example of Apple. The company's market capitalization is roughly $3 trillion and generates approximately $100 billion in free cash flow each year. I asserted that there was an important relationship between the two. Let me expand on this example and walk through the connection between free cash flow and valuation step by step.

Let's assume that Apple's 2024 free cash flow is $100 billion and let's assume that Apple is successful in growing its annual free cash flow 10 percent per year going forward for the next five years. Thus, we are assuming Apple has an annual free cash flow as represented in the following table:

Year	Free Cash Flow (in millions)
2024	100,000
2025	110,000
2026	121,000
2027	133,100
2028	146,410

How valuable are these free future cash flows in today's dollars? Well, they need to be discounted by some discount rate since future cash flow is less valuable than current cash flow. Because Apple is a very stable, high-quality company, the interest rate at which Apple can borrow money is very low. At the time of this writing, the U.S. government is borrowing money at an interest rate of 5 percent. Apple isn't as credit-worthy as the U.S. government, but it's about as good as it gets in the corporate world and so its cost of capital (that is, the cost to the company to raise capital) is just a few points above this figure. And Apple's cash flows are dependable and stable. Given all that, let's assume that the appropriate discount rate that we should apply to Apple's future cash flows is 8 percent. Then, we can calculate the present value of these future cash flows as follows:

Year	Free Cash Flow (in millions)	Present Value (in millions)
2024	100,000	100,000
2025	110,000	101,852
2026	121,000	103,738
2027	133,100	105,659
2028	146,410	107,616

Now we want to calculate the terminal value, which represents the sum of the estimated cash flows at the end of 2028, reflecting an expectation that Apple will grow essentially in perpetuity from that point forward. That sum then needs to be discounted back to the present value. To calculate terminal value, an assumption needs to be made about what the growth rate for the cash flows would look like in the future in perpetuity, known as the terminal growth rate. Let's assume that Apple's 10 percent rate can't last forever, so a more conservative growth rate might be 4 percent—just a bit faster than the 3 percent growth rate of global GDP over the last ten years.

The terminal value is calculated as the FCF of 2028 x (1 + terminal growth rate) / (discount rate - terminal growth rate). That is, $146,410 \times (1 + 4\%) / (8\% - 4\%) = \$3,806,660$. The present value of that sum, factoring in our 8% discount rate, is $2,798,009. Add that sum to the present value of the free cash flows from 2024–2028 and you get a total value of Apple—based on these assumptions—of $3,316,874. As of this writing, Apple's market capitalization is $3.32 trillion.

Year	Free Cash Flow (in millions)	Present Value (in millions)
2024	100,000	100,000
2025	110,000	101,852
2026	121,000	103,738
2027	133,100	105,659
2028	146,410	107,616
Terminal Value	3,806,660	2,798,009
Total		3,316,874

The Art and Science of Startup Valuations

Now that you understand the fundamentals of valuing a company, it is clear why estimating the future cash flow of a startup is fiendishly difficult. A few challenges:

1. Startups typically lose money every year. Thus, future cash flows are very difficult to estimate.

2. The cost of capital for a startup is much, much higher than that of an established public company. And the risks associated with future cash flows of a startup are very high. Thus, the discount rate varies widely.

3. The growth rate of startups can be astounding. Rather than 10–20 percent a year, startups can grow revenue and free cash flows over 100 percent per year. The higher growth rates suggest higher relative valuations, but there is also higher volatility to these growth rates.

4. The assumptions behind the terminal value feel absurd. How can you assume that a startup—with all its vulnerabilities and fragility—will be able to generate future cash flows in perpetuity? And yet for a company that will predictably lose money for the next few years and have modest profitability even ten years from now, nearly all the value sits in the terminal value.

The way most investors handle this challenge is that they use valuation multiples as a rule of thumb. A valuation multiple is a financial metric used to value a company by comparing it to similar companies or industry standards. Once you analyze and understand how Salesforce is valued as, say, a multiple of free cash flow or as a multiple of sales, then you can infer how other SaaS companies might be valued using similar ratios. Then you can compare the particular company you are evaluating to, say, Salesforce in terms of its growth rate, gross margins, competitive position, execution, and other subjective factors.

For example, at the time of this writing, Salesforce's price to sales or revenue multiple is 7x. The company's valuation is approximately $300 billion, and its revenue is $40 billion. Salesforce's free cash flow is roughly $11 billion and so its valuation as a multiple of free cash flow is roughly 27x. Note this free cash flow valuation multiple is lower than Apple's 30x, implying investors have more confidence in Apple's future cash flows than Salesforce's.

Revenue multiples are a popular shorthand for valuations of unprofitable startups. Many SaaS companies, like Salesforce, are very profitable and have gross margins of 75–80 percent and free cash flow margins of 30–40 percent. Therefore, SaaS companies carry high revenue multiples because there is an assumption that, at scale, they will have similar profitability profiles. The way the math works, a 30x free cash flow multiple valuation is the same as a 9x revenue valuation for a company that has free cash flow margins of 30 percent.

Valuations and Archetypes

In Chapter 6, I outlined a range of business model archetypes and asserted a series of revenue multiples for each of these valuations. Those assertions were based on looking at the best-in-class public market industry comparables. Here is a chart that summarizes my assertions:

Archetype	Revenue Multiples	Examples
SaaS	8 - 10x	Atlassian, Hubspot, MongoDB, Snowflake
Consumer subscription	6 - 8x	Life360, Netflix, Spotify
Advertising	3 - 5x	Facebook, Snap
Marketplace	3 - 5x	Booking.com, Etsy, Olo
Transaction-based	3 - 5x	Bill.com, Block, PayPal
Fintech/lending	3 - 5x	Affirm, SoFi
One time product sales	2 - 4x	Warby Parker, Wayfair

In each case, the range of valuation is impacted by the underlying characteristics and gross margin profile of each business model. For example, as noted earlier, SaaS business models tend to have 75–80 percent gross margins and so incremental revenue is highly accretive, and these companies can become very profitable at scale—as much as 30–40 percent in free cash flow margins. E-commerce product sales businesses tend to have lower gross margins and lower overall profit margins. For example, Wayfair earned $12 billion in revenue in 2023 with a 30 percent gross margin and a free cash flow margin of less than 10 percent.

Note that these ranges and examples represent the best-in-class for these categories. There are always a few outliers—sometimes for brief moments in time—because the market believes fervently in the potential for their future cash flows to exceed even the best-in-class companies. That's why a company like NVIDIA might achieve a valuation of 20–25x revenue—because there is a belief that their future free cash flows are going to grow very rapidly, as evidenced by the fact that their revenue is growing at an astonishing 200 percent year over year.

Forward Multiples

Although NVIDIA is an outlier among public companies, good startups can grow very quickly. A VC friend of mine, Neeraj Agrawal, coined a term in a blog post that many investors now use as shorthand: triple, triple, double, double.[67] The notion was that the most successful young companies can and should grow so fast that after they reach a critical mass point of $1 or $2 million in revenue, they should be tripling their revenue for two years in a row and then doubling their revenue for two years after that on their path to success.

This pattern of rapid growth is why many investors talk about "forward multiples" when investing in startups rather than current revenue multiples. For example, when evaluating a $20 million revenue company that is forecasted to triple in the next year, an investor might conclude that rather than invest at a valuation of, say, 10x current revenue—or $200 million—they might be able to justify investing at some multiple of future revenue. If an investor concludes that there is a high likelihood that the company will indeed triple to $60 million in revenue in a year, they might be willing to pay a modest discount to the 10x multiple rule of thumb—for example, 6x or even 8x forward revenue or a valuation of $360 million or $480 million. This analysis methodology is known as "forward multiples" and has grown in popularity among investors.

For a company that is bringing in a mere $1 million in revenue but has a good shot at achieving $3 million in revenue in a year, they might be able to convince investors that they are worth a $25–30 million valuation. If a VC firm believes the startup has a good shot at achieving $10 million in revenue in a year, they might be willing to pay closer to $80–100 million—even if that valuation represents 80–100x current revenue. I told you in Chapter 4. As VC firm A16Z put it in a blog post on this topic, "As growth investors who believe in companies executing big, risky visions, we look at how long a company could grow at a high rate when assessing potential investments. This growth could be a function of product differentiation, go-to-market operations, sheet market size, new geographies, and expansion into adjacent categories."[68]

Valuations in private companies are struck differently than in public companies—and that difference is a critical one for founders to understand. In the public markets, the stock of a public company is very liquid. It can be bought or sold at any given moment by thousands and thousands of potential investors. Thus, the market for the price of a public stock is very efficient and cannot typically be influenced by one individual or institution.

There is no liquid market for the stock of a private company. The price is not set by an efficient market of buyers and sellers every minute of every day. Instead, it is set once every few years by a few institutional venture capitalists. It only takes one VC to set the price of a private company. And when that price is set, the transaction is struck. Thus, private companies are often inherently mispriced. The private market is an inefficient part of the market, allowing different investors with different views of the prospects—in particular, the future cash flows—of a company to hold sway. If one investor is very bullish about the future cash flows of a startup, they might have a bullish view on their forward prospects and therefore be willing to pay a healthy forward multiple.

Now you can see why Khatabook, featured in Chapter 5, was able to secure a funding round at a $600 million valuation despite having almost zero revenue at the time. The investors who funded that round—and it only took a few—convinced themselves that the company's rapid user growth and strong network effects would eventually lead to a powerful monetization and free cash flow engine. When they did the future cash flow and terminal value math, their analysis got them to believe that $600 million could be justified because they had high conviction that the future value of the company would be in the billions. Whether they will be proven right remains to be seen. The investors who put money in WeWork at a $47 billion valuation lost all their money while those who invested in Facebook (a.k.a., Meta) at a roughly similar valuation made a killing.

Later-stage investors—also known as growth investors—typically aim to achieve a 3–5x return on their investment to justify the high risk they are taking. If they can achieve this return in, say, 4–6 years, they can earn an annual investment return of 20–50 percent (in other words, a 3x return in six years represents a 20 percent annual return and 5x return in four years represents a 50 percent annual return). Even if they have a few bad investments, a few good ones matching this profile result in a strong performing fund.

Valuations in the Beginning

With all that as background, let's return to focus on the founder who is just getting started. How does she value her company when she has zero revenue and no customers?

The investor who is investing in that nascent, early-stage startup is asking themselves the following questions:

1. What is the inherent business model underlying this startup and how valuable will it be if it succeeds? That is, if everything works out amazingly well, how big could this be and how valuable could it be based on the business model fundamentals, the relevant

set of comparable companies with a similar business model, and how those comparable companies are valued?[69]

2. What is the likelihood of achieving the outsized success imagined above?

3. When this company is next raising money, let's say in two to three years, how valuable will they be? Will I get "paid" for the risk I'm taking now, or should I just wait and see since the price of the company's shares will likely not be much more expensive the next time they raise money?

When early-stage investors like Flybridge analyze startup valuations, we have a more ambitious outcome that we are aiming for because we are expecting more risk. In other words, the answer to question 2 above (likelihood of success) is naturally going to be lower for a younger company just getting off the ground than an established leader who has achieved product-market fit and is exhibiting a business model that is starting to hum.

The rule of thumb that early-stage investors try to follow is to achieve a 10–20x return on their investments. Why so high? Because whereas later-stage investors may expect one out of two or three investments to achieve their target returns, early-stage investors expect to achieve these kinds of outsized returns only 5–10 percent of the time. And early-stage investors expect to suffer a great deal of dilution in the ensuing years while they are holding on to their shares. In a portfolio of forty to fifty companies, the best VC funds might have two to three companies that can achieve these kinds of returns.

Another rule of thumb that most early-stage founders follow is to avoid selling more than 20 percent of their company in any given round of financing. I call that "the rule of 20." Because there are likely to be multiple rounds of financing ahead, it is prudent to avoid diluting yourself too much in that first round. Further, if things go well, that first round of financing is going to be your most expensive round as a founder. So be protective of your equity!

Let's take an example. Assume a founder comes to us and says they want to raise $1 million in a pre-seed round. Let's say she wants to value her startup at a $10 million post-money (the pre-money valuation would be $9 million in this example—$1 million of investment plus $9 million pre-money valuation equals a $10 million post-money valuation). The logic we follow to convince ourselves to get to yes is as follows:

- In 18–24 months, when the company needs to raise money again, will they have achieved enough to justify raising money at a valuation 2–3x their current valuation of $10 million? In other words, do we believe they can be worth $20–30 million in 18–24 months?

- When the dust settles and the company finally exits (through either an IPO or an

M&A acquisition), can the company be worth 10–20x? Factoring in future dilution, 20–30x? In other words, can this company eventually be worth at least $200–300 million?

If the founder pushes us to pay $20 million post-money, then we must double all of these figures.

Actual Startup Valuation Figures

The firm Carta provides an excellent quarterly report that shares the "state of the private markets" and various valuation comparables. Looking at the Q2 2024 report, you can see the following valuations by stage:

Stage	Median Pre-Money Valuation	Median Financing Size	Implied Dilution
Seed	$14.8M	$3M	16.9%
Series A	$42.2M (2.8x previous round)	$9M	17.6%
Series B	$116.6M (2.8x)	$19.2M	14.1%
Series C	$200M (1.7x)	$28M	12.2%

The following rules of thumb are clear: Ideally, in each round of financing, founders are selling less than 20 percent of their company to investors and raising capital at a valuation that is 2–3x their previous round.

In practice, first-round valuations are based on subjective elements such as

1. The track record of the founder: Have they been an executive at a large valuable company in the past?

2. The importance and size of the opportunity: Is the startup pursuing a niche or a massive market?

3. What is the depth of the technical IP and what is the competitive moat?

4. How competitive the deal dynamics are: Startup valuations are a simple case of supply and demand

Not all startups follow the median path. And not all startups are able to "clear the bar" and achieve funding at all. But hopefully this appendix arms you as a founder with the knowledge of how startup valuations work and gives you a few hints at how to position your startup to be more attractive to investors.

For a deeper dive into how to raise startup financing, you might want to read my book, *Mastering the VC Game*.

Appendix C

Cohort Analysis Examples and Explanation

WHILE GROWTH OFTEN GARNERS the most attention, a company's ability to retain users over time is a more telling indicator of long-term success. Cohort analysis is an essential tool for understanding the dynamics of user behavior over time, making it invaluable for startups aiming to achieve and maintain product-market fit. By grouping users based on a common characteristic or event—such as the week or month they first started using a product—cohort analysis allows founders to track how different segments of users engage with the product over various time periods. This analysis provides critical insights into retention rates, user engagement, and the overall effectiveness of product iterations and marketing strategies. Unlike traditional metrics that provide a snapshot view, cohort analysis offers a longitudinal perspective, revealing trends and patterns that are crucial for informed decision-making.

For both founders and joiners, the importance of cohort analysis cannot be overstated. As startups iterate on their products and go-to-market strategies, they need to understand not just if their user base is growing, but how sustainable that growth is. Cohort analysis helps identify whether improvements are genuinely enhancing user experience and retention or if they're merely short-term gains. It also highlights potential issues such as high churn rates among specific user groups, allowing for targeted interventions. In essence, cohort analysis equips startups with the data-driven insights needed to refine their offerings and scale effectively.

Potential issues such as high churn among specific user groups can be particularly detrimental if not identified and addressed promptly. High churn rates indicate that users are not finding the product valuable enough to continue using it, which can undermine growth efforts and waste valuable resources on acquiring new users who will not stick around (a.k.a., "the leaky bucket problem"). For instance, if a startup notices that retention rates drop significantly after the first month, it may need to improve onboarding processes, enhance the core features, or provide better customer support. By conducting a detailed cohort analysis, startups can pinpoint the exact stage at which users disengage and take targeted actions to improve retention. This approach not only helps in refining the product but also in demonstrating to investors that the startup is data-driven and committed to continuous improvement.

Getting Started

Creating a cohort analysis begins with defining the specific cohorts you want to track. A cohort typically refers to a group of customers or users who share a common characteristic. For example, capturing all customers within a defined period such as signing up for a product in the same month. Another example would be to organize cohorts by type of user—for example, capturing all customers segmented by their number of employees.

The first step is to collect relevant data on your users, which often includes sign-up dates, engagement metrics, and key performance indicators like purchase behavior or retention rates. Ensure that the data collected is both granular and timestamped to allow for the accurate segmentation of users into cohorts. It is critical to establish a consistent definition of "activation" or the point at which a user is considered "active" or "engaged"—this could be their first download, sign-in, purchase, a certain level of interaction, or another meaningful metric aligned with your business objectives.

Normalization is essential to making your cohort data comparable across different timeframes. To do this, you must account for any variations in data collection methods, differing cohort sizes, and external factors that might influence behavior. For example, if marketing campaigns or product releases vary significantly across cohorts, consider adjusting the data to ensure you're making an apples-to-apples comparison. After this, you'll structure the data into a matrix, with time intervals (weeks, months, etc.) on one axis and cohorts on the other. The final step is calculating and visualizing key metrics, such as retention rates or revenue per cohort, across time periods to reveal trends and actionable insights.

Example: Squire

Let's take the Squire case study as an example. The Squire team carefully tracked the average revenue per user (ARPU). From this data, they were able to chart out a series of cohorts of users and how their ARPU evolved each month. Those figures are shown below from their first thirteen months of operation.

Normalizing the data set and anchoring each cohort to a "time zero" is an important step in the analysis as it allows for like-to-like comparisons.

There are two ways to view a cohort chart like this. First, scan each column horizontally to get a sense of how the cohorts are changing over time. For example, the first highlighted cohort below begins at month zero with an ARPU of $157 (note: I often recommend ignoring the first cohort as the sample size can be too small to be significant). By month twelve (the last available month of data for this cohort), that figure grows to $503—a growth of 220 percent. The second highlighted cohort below begins at month zero with an ARPU of $223. By month eight (the last available month of data for this cohort), that figure grows to $783 — a growth of 251 percent over an even shorter time period. In other words, customers are growing in their usage of the Squire product

ARPU in Month Numbers Since Joining Squire (January 2019 – February 2020)

0	1	2	3	4	5	6	7	8	9	10	11	12	13
$339	$327	$453	$512	$649	$622	$560	$671	$529	$615	$572	$593	$587	$600
157	292	273	442	427	424	476	452	482	452	502	487	503	
150	284	327	354	327	342	349	465	434	355	503	569		
195	255	304	335	418	369	499	499	494	516	497			
263	432	512	609	568	641	615	660	683	793				
223	370	547	552	812	791	881	782	783					
179	481	473	576	543	596	543	579						
531	904	395	458	468	473	494							
209	266	282	349	349	382								
162	322	356	367	389									
200	370	383	402										
320	346	395											
241	274												
230													

ARPU in Month Numbers Since Joining Squire (January 2019 – February 2020)

0	1	2	3	4	5	6	7	8	9	10	11	12	13
$339	$327	$453	$512	$649	$622	$560	$671	$529	$615	$572	$593	$587	$600
157	292	273	442	427	424	476	452	482	452	502	487	503	
150	284	327	354	327	342	349	465	434	355	503	569		
195	255	304	335	418	369	499	499	494	516	497			
263	432	512	609	568	641	615	660	683	793				
223	370	547	552	812	791	881	782	783					
179	481	473	576	543	596	543	579						
531	904	395	458	468	473	494							
209	266	282	349	349	382								
162	322	356	367	389									
200	370	383	402										
320	346	395											
241	274												
230													

over time, resulting in growing revenue on a per-customer basis for the company. A good indicator of product-market fit is this pattern of growth, known as negative churn. Negative churn occurs when the revenue gained from existing customers, through upselling or increasing usage, exceeds the lost revenue from churned customers.

The other way to view a cohort chart like this is to scan each column vertically to get a sense of how the cohorts are changing over time. For example, the first highlighted column shows that the second cohort began month one with an ARPU of $327. A number of months later, the latest cohort starts month one with an ARPU of $274—a decrease of 16 percent. That might be worrisome as it shows that as the cohorts get larger over time, the quality may not be as high. If we ignore the first cohort (again, due to small sample size) and look at the second cohort, it began month one with an ARPU of $292 and then dropped to $274 in the later cohort—not as much of a decrease (6 percent) and so perhaps more encouraging.

But then the second highlighted column shows a more encouraging picture. The change in cohort behavior in month eight, after customers have a chance to experience the product for a while. If we again ignore the first cohort and look at the second cohort, we see it began month eight with an ARPU of $482. A few months later, the sixth cohort reaches month 8 with an ARPU of $783—an improvement of 62 percent. That is very encouraging!

So now we have a very interesting pattern. Squire's customers are not only using the product more over time (ARPU growing horizontally on this chart), but each new cohort of

ARPU in Month Numbers Since Joining Squire (January 2019 – February 2020)

0	1	2	3	4	5	6	7	8	9	10	11	12	13
$339	$327	$453	$512	$649	$622	$560	$671	$529	$615	$572	$593	$587	$600
157	292	273	442	427	424	476	452	482	452	502	487	503	
150	284	327	354	327	342	349	465	434	355	503	569		
195	255	304	335	418	369	499	499	494	516	497			
263	432	512	609	568	641	615	660	683	793				
223	370	547	552	812	791	881	782	783					
179	481	473	576	543	596	543	579						
531	904	395	458	468	473	494							
209	266	282	349	349	382								
162	322	356	367	389									
200	370	383	402										
320	346	395											
241	274												
230													

customers is improving in their usage of the product from the beginning of their experience and then throughout (ARPU growing vertically). That is a sign of strong product-market fit!

Another technique for deriving insights from cohort analysis is to chart the results in a graph. Below we chart the Squire ARPU cohorts over time.

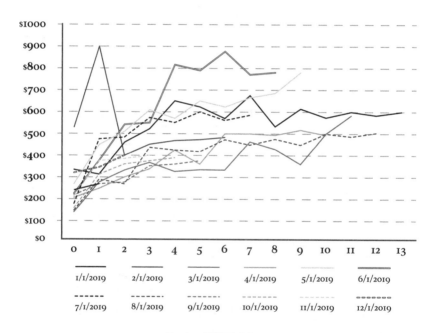

Squire ARPU Cohorts

Cohort charts can be complicated if there is a lot of data—which is the case here even though we are only showing one year's worth of cohorts. But if you take away some of the noise, you can see the pattern of cohorts beginning in the $200–$300 ARPU range and rapidly growing to the $600–$800 ARPU range after only six to eight months.

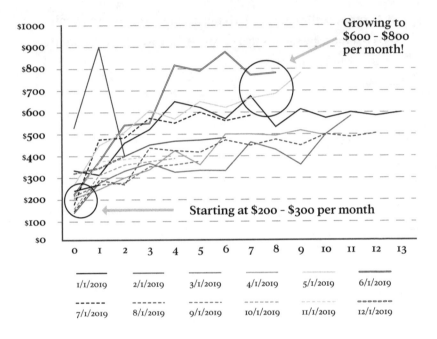

Squire ARPU Cohorts

Venture capitalist Fred Wilson wrote on this topic: "Entrepreneurs always ask what the one number they should focus on for raising money. I always say, 'ninety-day retention numbers for your acquisition cohorts.' There's a common view . . . that growth is the one thing you should focus on. But it's hard to grow if you are churning your users. And if you are paying for user acquisition, as many startups do in search of growth, then retention/churn becomes even more important."[70]

It should be acknowledged that cohort data in startups is initially messy. Small data sets, idiosyncratic factors, strategy or product shifts, and the challenges of data coming from immature systems can complicate your analysis and obscure clear insights. Therefore, cohort analysis for early-stage startups should be done with a mix of humility, practical sense, and business judgment. That early messiness is a part of the startup journey—and all the more reason to take an iterative, experimentation approach.

Appendix D

a16z's AI Apps Unwrapped

AT THE END OF 2024, VC firm Andreessen Horowitz (a16z) released a list of their favorite AI apps and tools.[71] The list is well-curated and will be relevant for at least a few years, so I wanted to share with you here:

General Assistance

- *Perplexity* – AI-powered search engine and research assistant
- *Claude (Anthropic)* – general chatbot, great for projects and sharing work
- *ChatGPT* – you know this one, but check out Advanced Voice Mode to talk to AI

Get Work Done

- *Granola* – AI notetaker that listens to your meetings and formats the transcript into notes
- *Wispr Flow* – AI voice dictation that turns your speech into text in any app
- *Gamma* – make decks, docs, and websites to present your ideas with AI
- *Adobe* – Summarize and chat with PDFs
- *Cubby* – AI workspace built for collaborative research
- *Cora* – AI email assistant that organizes your inbox and automates responses
- *Lindy* – Build AI agents to automate your workflows

Build an Audience

- *Delphi* – AI text, voice, and video clones to chat with your audience
- *HeyGen* – AI avatars to scale your content production or translate your videos
- *Argil* – AI avatars for social media videos
- *Overlap, Opus* – Turn your long-form videos into short viral clips with AI
- *Persona* – AI agent builder for creators
- *Captions* – AI avatars and video editing (e.g. auto-captions, correct eye contact)

Build a Product

- *Cursor* – AI code editor that knows your codebase
- *Replit* – AI agents to make apps and sites from natural language
- *Anychat* – Use any AI model in one place
- *Codeium* – AI-powered autocomplete for your code

Get Creative

- *ElevenLabs* – Realistic AI voices
- *Suno, Udio* – Create songs and music from text prompts
- *Midjourney, Ideogram, Playground* – AI image generation
- *Runway, Kling, Viggle* – AI video generation
- *Krea* – AI creative canvas to make and enhance images and video
- *Photoroom* – AI image editor, great for product photos and visuals

Learn or Grow

- ***Rosebud*** – Interactive journal that uses AI to surface insights
- ***Good Inside*** – Parenting co-pilot with personalized support
- ***Ada Health*** – Get an AI-powered assessment of medical symptoms
- ***Ash*** – Personalized AI counselor / coach
- ***NotebookLM*** – Turn any document into an AI podcast
- ***Particle*** – AI news app that combines multiple articles into summarized stories

Have Fun

- ***Remix*** – Social app for creating and sharing AI images and video
- ***Meta Imagine*** – Make AI images of yourself, family, and friends in Meta apps
- ***Grok*** – Chatbot from xAI
- ***Curio*** – Toys for kids to talk to, powered by AI voices

Endnotes

1. A term I coined that refers to the world of startups, and the subject of my second book, *Entering Startupland*.

2. Bussgang, Jeffrey J., Zoë B. Cullen, William R. Kerr, Benjamin N. Roth, and Michael Norris. "A Close Shave at Squire." Harvard Business School Case 821-073, July 2021.

3. Gounares, Alexander. "The Billion Dollar, Single Person Company." ThoughtfulBits: Ideas That Matter (blog), May 20, 2024. https://www.thoughtfulbits.me/p/the-billion-dollar-single-person.

4. "The Era of Scaling Without Growing & the Meaning Economy." *Implications, by Scott Belsky* (blog), June 23, 2024. https://www.implications.com/p/the-era-of-scaling-without-growing.

5. "The Productivity Paradox of Information Technology: Review and Assessment," n.d. http://ccs.mit.edu/papers/CCSWP130/ccswp130.html.

6. Du Preez, Derek. "ServiceNow CEO Says Process Optimization Is the 'Single Biggest Generative AI Use Case in the World Today.'" *Diginomica*, April 25, 2024. https://diginomica.com/servicenow-ceo-says-process-optimization-single-biggest-generative-ai-use-case-world-today.

7. Barbaro, M., Metz, C., Tan, S., Johnson, M. S., Zadie, M., Novetsky, R., Georges, M., Baylen, L. O., Wong, D., Powell, D., McCusker, P., & Wood, C. (2024, April 16). A.I.'s original sin. *The New York Times*. https://www.nytimes.com/2024/04/16/podcasts/the-daily/ai-data.html?

8   Shaping Europe's Digital Future. "AI Act," December 12, 2024. https://digital-strategy.ec.europa.eu/en/policies/regulatory-framework-ai.

9   Heikkilä, Melissa. "Making an Image With Generative AI Uses as Much Energy as Charging Your Phone." *MIT Technology Review*, December 1, 2023. https://www.technologyreview.com/2023/12/01/1084189/making-an-image-with-generative-ai-uses-as-much-energy-as-charging-your-phone/.

10 IEA. "Executive Summary – Electricity 2024 – Analysis - IEA," n.d. https://www.iea.org/reports/electricity-2024/executive-summary.

11 Situational Awareness - "The Decade Ahead," June 6, 2024. https://situational-awareness.ai/.

12 Bussgang, Jeffrey J., and Olivia Hull. "Classtivity: Payal's Pirouette." Harvard Business School Case 817-002, January 2017. (Revised October 2023.)

13 Crook, Jordan. "ClassPass Lands $12 Million Series a Led by Fritz Lanman, Hank Vigil." *TechCrunch*, September 17, 2014. https://techcrunch.com/2014/09/17/classpass-lands-12-million-series-a-led-by-fritz-lanman-hank-vigil/.

14 Rajesh, Ananya Mariam. "Mindbody ClassPass to Go Public in 12-18 Months, CEO Says." *Reuters*, August 14, 2024. https://www.reuters.com/markets/deals/mindbody-classpass-go-public-12-18-months-ceo-says-2024-08-14/.

15 "The 4 Levels of PMF," n.d. https://pmf.firstround.com/levels.

16 Staff, First Round. "How Superhuman Built an Engine to Find Product Market Fit." First Round Review, October 24, 2024.

17 Konrad, Alex. "Superhuman Raises $75 Million for Its Waitlist-Only Email Productivity App." *Forbes*, August 4, 2021. https://www.forbes.com/sites/alexkonrad/2021/08/04/superhuman-raises-75-million-for-its-waitlist-only-email-productivity-app/.

18 Field, Hayden. "OpenAI's Active User Count Soars to 300 Million People per Week." CNBC, December 4, 2024. https://www.cnbc.com/2024/12/04/openais-active-user-count-soars-to-300-million-people-per-week.html.

19 XM Institute. "Calibrating NPS Across 18 Countries | XM Institute," January 11, 2023. https://www.xminstitute.com/research/calibrating-nps-18-countries/.

20 Bussgang, Jeffrey J., and Olivia Hull. "C16 Biosciences: Lab-Grown Palm Oil." Harvard Business School Case 820-008, October 2019. (Revised November 2019.)

21 Girotra, Karan, Lennart Meincke, Christian Terwiesch, and Karl T. Ulrich. "Ideas Are Dimes a Dozen: Large Language Models for Idea Generation in Innovation." *SSRN Electronic Journal*, January 1, 202https://doi.org/10.2139/ssrn.4526071.

22 Bussgang, Jeffrey J., and Mel Martin. "AllSpice: GitHub for Hardware Engineers." Harvard Business School Case 823-022, September 2022.

23 The main takeaway from Moore's research is the dangerous *chasm* between the innovator/early adopter markets and the majority markets. The challenge for every growing business is crossing the chasm and into the mainstream. This scaling challenge is covered in Chapter 8.

24 Ashish Vaswani, Noam Shazeer, Niki Parmar, et al., "Attention Is All You Need," arXiv.org, June 12, 2017, https://arxiv.org/abs/1706.03762.

25 Jeffrey Bussgang and Oliver Jay. "How Software Companies Can Avoid the Trap of Product-Led Growth," April 18, 2024. https://hbr.org/2023/09/how-software-companies-can-avoid-the-trap-of-product-led-growth.

26 Bussgang, Jeffrey J., Jeffrey F. Rayport, and Olivia Hull. "Delivering the Goods at Shippo." Harvard Business School Case 817-065, January 2017. (Revised October 2021.)

27 Bussgang, Jeffrey J., and Julia Kelley. "Giving Birth to Ovia Health." Harvard Business School Case 818-004, January 2018. (Revised September 2023.)

28 The term *10x Club* was originally coined by venture capitalist Bill Gurley of Benchmark.

29 Bussgang, Jeffrey J., Allison H. Mnookin, and James Barnett. "KhataBook." Harvard Business School Case 821-006, November 2020. (Revised September 2021.)

30. Christine Deakers, "State of the Cloud 2023," Bessemer Venture Partners, October 31, 2023, https://www.bvp.com/atlas/state-of-the-cloud-2023.

31. Bill Gurley's more detailed post on this topic can be found on his excellent blog, *Above the Crowd*.

32. Bussgang, Jeffrey J., and Julia Kelley. "Giving Birth to Ovia Health." Harvard Business School Case 818-004, January 2018. (Revised September 2023.)

33. Harvard Business School. "Want to Be an Entrepreneur?," March 19, 2001. https://www.library.hbs.edu/working-knowledge/want-to-be-an-entrepreneur-part-i.

34. Eisenmann, Tom. "Why Start-Ups Fail," September 17, 2021. https://hbr.org/2021/05/why-start-ups-fail.

35. "How Much More Efficient Should a SaaS Startup Be When Using AI? By @ttunguz," n.d. https://tomtunguz.com/how-much-more-profitable-saas/.

36. How many countless hours of our lives did we collectively lose to unsaved work on Word? It pains me to think about it.

37. Kelly, Jack. "AI Writes Over 25% of Code at Google—What Does the Future Look Like for Software Engineers?" *Forbes*, November 4, 2024. https://www.forbes.com/sites/jackkelly/2024/11/01/ai-code-and-the-future-of-software-engineers/.

38. Zheyuan Cui et al., "The Effects of Generative AI on High Skilled Work: Evidence from Three Field Experiments with Software Developers," *SSRN*, September 5, 2024, https://doi.org/10.2139/ssrn.4945566.

39. Bussgang, Jeffrey J., Jeffrey F. Rayport, and Olivia Hull. "Delivering the Goods at Shippo." Harvard Business School Case 817-065, January 2017. (Revised October 2021.)

40. "How Shippo Found Product-Market Fit," n.d. https://www.unusual.vc/post/how-shippo-found-product-market-fit.

41. Fowler, Martin. "Bliki: Technical Debt Quadrant." martinfowler.com, n.d. https://martinfowler.com/bliki/TechnicalDebtQuadrant.html.

42 Leslie, Mark. "The Sales Learning Curve." *Harvard Business Review*, August 1, 2014. https://hbr.org/2006/07/the-sales-learning-curve.

43 "How Much More Efficient Should a SaaS Startup Be When Using AI? By @Ttunguz."

44 Two good books that expand on this topic are *Predictable Revenue: Turn Your Business into a Sales Machine with the $100 Million Best Practices of Salesforce.com* by Aaron Ross and Marylou Tyler, and *The Sales Acceleration Formula: Using Data, Technology, and Inbound Selling to go from $0 to $100 Million* by Mark Roberge.

45 "How Much More Efficient Should a SaaS Startup Be When Using AI? By @Ttunguz."

46 Bussgang, Jeffrey J., Bonnie Yining Cao, and Dawn H. Lau. "Sprout Solutions." Harvard Business School Case 824-052, October 2023. (Revised November 2023.)

47 "When Community Becomes Your Competitive Advantage." *Harvard Business Review*, 21 January 2020 by Bussgang, Jeffrey and Jono Bacon.

48 Jeff Bussgang, "Hey Graduates: Forget Plastics – It's All About Machine Learning," SEEING BOTH SIDES, May 11, 2012, https://seeingbothsides.com/2012/05/11/forget-plastics-its-all-about-machine-learning/.

49 This is an expansion of the term "startup joiner," which I coined when I wrote *Entering Startupland: An Essential Guide to Finding the Right Job* to focus attention on employees 2 through 2,000 at a startup.

50 Mollick, Ethan. "Reshaping the Tree: Rebuilding Organizations for AI." *One Useful Thing* (blog), November 27, 2023, https://www.oneusefulthing.org/p/reshaping-the-tree-rebuilding-organizations.

51 Robert M. Pirsig, Zen And the Art of Motorcycle Maintenance: An Inquiry Into Values, 1974, http://ci.nii.ac.jp/ncid/BA83974814.

52 If titles are given out, I prefer "head of product" or "head of sales" rather than hierarchical titles like Vice President because it gives more flexibility for their roles to change over time.

53 Reid Hoffman and Chris Yeh, *Blitzscaling: The Lightning-Fast Path to Building Massively Valuable Companies* (Crown Currency, 2018).

54 Goldman Sachs. "AI Is Poised to Drive 160% Increase in Data Center Power Demand," May 14, 2024. Goldman Sachs. https://www.goldmansachs.com/insights/articles/AI-poised-to-drive-160-increase-in-power-demand.

55 Belanger, Ashley, and Ashley Belanger. "OpenAI Asked U.S. to Approve Energy-Guzzling 5GW Data Centers, Report Says." *Ars Technica*, September 25, 2024. https://arstechnica.com/tech-policy/2024/09/openai-asked-us-to-approve-energy-guzzling-5gw-data-centers-report-says/.

56 Harvard Law Review. "NYT V. OpenAI: The Times's About-Face." Harvard Law Review, April 10, 2024. https://harvardlawreview.org/blog/2024/04/nyt-v-openai-the-timess-about-face/.

57 Sherrer, Kara. "HarperCollins Book Catalog to Train Microsoft AI Models for the Next 3 Years." *eWEEK*, December 4, 2024. https://www.eweek.com/news/harpercollins-books-train-microsoft-ai-models/.

58 "The Vector Database to Build Knowledgeable AI | Pinecone," n.d. https://www.pinecone.io/.

59 Ducharme, Jamie. "How Juul Got Vaporized." *TIME*, May 17, 2021. https://time.com/6048234/juul-downfall/.

60 Ducharme, "How Juul Got Vaporized."

61 "Wisconsin DOJ and 32 Other States Finalize $435 Million Agreement with JUUL Labs | Wisconsin Department of Justice," n.d. https://www.doj.state.wi.us/news-releases/wisconsin-doj-and-32-other-states-finalize-435-million-agreement-juul-labs.

62 For a full breakdown of the Theranos debacle, see O'Brien, Sara Ashley. "The Rise and Fall of Theranos: A Timeline." *CNN*, July 7, 2022. https://www.cnn.com/2022/07/07/tech/theranos-rise-and-fall/index.html.

63 University of California Press. "Behind the Startup by Benjamin Shestakofsky," n.d. https://www.ucpress.edu/books/behind-the-startup/.

64 Baboolall, David, Kelemwork Cook, Nick Noel, Shelley Stewart, and Nina Yancy. "Building Supportive Ecosystems for Black-owned U.S. Businesses." McKinsey & Company, October 29, 2020. https://www.mckinsey.com/industries/public-sector/our-insights/building-supportive-ecosystems-for-black-owned-us-businesses.

65 Responsible Innovation Labs, "Responsible AI," n.d., https://www.rilabs.org/responsible-ai.

66 James Clear, *Atomic Habits: An Easy & Proven Way to Build Good Habits & Break Bad Ones*, 2018, https://catalog.umj.ac.id/index.php?p=show_detail&id=62390.

67 Battery Ventures. "Helping Entrepreneurs "Triple, Triple, Double, Double, Double" to a Billion-Dollar Company." Battery Ventures, October 24, 2024. https://www.battery.com/blog/helping-entrepreneurs-triple-triple-double-double-double-to-a-billion-dollar-company/.

68 Immerman, Alex, David George, Alex Immerman, and David George. "When Entry Multiples Don't Matter." Andreessen Horowitz, September 3, 2023. https://a16z.com/when-entry-multiples-dont-matter/.

69 Teng, Janelle. "Growth-Stage Startup Valuations: Guiding Principles for Comps Selection." *Next Big Teng* (blog), June 16, 2023. https://nextbigteng.substack.com/p/valuations-guiding-principles-comps-selection.

70 Wilson, Fred. "Growth Vs Retention - AVC." AVC, July 27, 2015. https://avc.com/2015/07/growth-vs-retention/.

71 Andreessen Horowitz, "Apps Unwrapped | Andreessen Horowitz," December 17, 2024, https://a16z.com/apps-unwrapped/.

Acknowledgements

THIS BOOK WAS CREATED with the support of a wide range of teammates, colleagues, and friends. It sprung from a conversation in late 2022 with Jermey Matthews, then a senior editor at The MIT Press and later an editor at McKinsey & Company. Jermey encouraged me to write a book based on my popular HBS course, Launching Tech Ventures (LTV). While we were discussing this idea, the AI revolution was exploding all around us, and the two threads came together with Jermey's capable guidance and input. Jermey connected me with Damn Gravity's Ben Putano, an entrepreneurial, creative publisher and editor with boundless enthusiasm for startups and innovation. Jermey and Ben were teammates and thought partners throughout, helping create what you have in your hands. If books could have co-founders, Jermey and Ben would be mine.

My Flybridge colleagues have been invaluable in my learning journey and many of the lessons here, as well as the case studies, emerged from our joint work. I want to particularly thank David Aronoff, Chip Hazard, and Jesse Middleton who have been cherished long-term partners over many decades. I also deeply appreciate the hundreds of founders in the Flybridge portfolio who have invited us onto their capitalization table and taught me so much about entrepreneurship. Many of them were kind enough to allow me to feature them in this book.

Speaking of learning, my HBS colleagues have been incredible teachers and collaborators throughout this journey. My LTV course colleagues have particularly helped shape my thinking over nearly 15 years of teaching—including Tom Eisenmann, Jeffrey Rayport, Sam Clemens, Reza Satchu, Lindsay Hyde, and Christina Wallace. And the over 2000 students who have taken the course are the reason I teach and the reason this book exists. Thank you for all you have taught me along the way.

During the writing journey, Ben had the brilliant idea of creating a cohort of Alpha Readers—150 strong—who would provide input along the way. They read each chapter upon completion and tore it apart, providing critical yet helpful guidance. I am grateful to others who provided detailed feedback on drafts of varying maturity, including Marion Annau, Shai Bernstein, Zach Brauer, Gonzalo Cervantes, Gary Eichhorn, Tom Eisenmann, Funke Faweya, Walter Frick, John Gannon, Giorgos Geroukos, Kofi HairRalston, Debi Kleiman, Andrew MacDowell, Julia Maltby, Ben Parfitt, Daniel Porras Reyes, Jessica Rosenbloom, Kenny and Sean Salas, Ayushi Sinha, Tim Sullivan, Mitch Weiss, and Laura Whitmer.

Finally, my wife Lynda and kids—Josh, Jackie, JJ, and Jonah—remain the center of my universe. This book is a tribute to their love and support.

About the Author

JEFFREY J. BUSSGANG HAS spent his career at the intersection of startup experimentation and transformative technology. For over a decade, he has taught one of Harvard Business School's most popular entrepreneurship courses, Launching Tech Ventures, where he's helped thousands of founders build and scale successful companies.

As co-founder of Flybridge Capital Partners (a leading seed VC fund with over $1 billion under management), he and his partners were early to recognize AI's potential to transform startups, evolving the firm's investment thesis to focus exclusively on AI-forward companies and founders.

Before his time as an educator and investor, Bussgang experienced the startup journey firsthand—joining Open Market's early team and helping lead it through an IPO in 1996, and then co-founding Upromise in 2000, which was successfully acquired.

This unique vantage point—as professor, investor, and startup executive—has given Bussgang an unmatched perspective on how AI is revolutionizing the startup landscape.

Now, he's sharing the frameworks and insights to help a new generation of founders build faster and smarter than ever before.

To read more from Jeffrey Bussgang, pick up his other books, *Mastering the VC Game* and *Entering Startupland*. You can also follow him on LinkedIn and read his blog (also called The Experimentation Machine).